OUTRAGE MACHINE

OUTRAGE MACHINE

*How Tech Amplifies Discontent,
Disrupts Democracy—And What We Can Do
About It*

TOBIAS ROSE-STOCKWELL

LEGACY
LIT

New York   Boston

Legacy Lit, an imprint of Hachette Books
Hachette Book Group
1290 Avenue of the Americas
New York, NY 10104
LegacyLitBooks.com
Twitter.com/LegacyLitBooks
Instagram.com/LegacyLitBooks

First Edition: July 2023

Grand Central Publishing is a division of Hachette Book Group, Inc. The Legacy Lit and Grand Central Publishing names and logos are trademarks of Hachette Book Group, Inc.

The Hachette Speakers Bureau provides a wide range of authors for speaking events. To find out more, go to hachettespeakersbureau.com or email HachetteSpeakers@hbgusa.com.

The publisher is not responsible for websites (or their content) that are not owned by the publisher.

Library of Congress Control Number: 2023003954

ISBNs: 9780306923326 (hardcover); 9780306923319 (ebook)

Printed in the United States of America

LSC-C

Printing 1, 2023

For Ani.

Turning and turning in the widening gyre
The falcon cannot hear the falconer;
Things fall apart; the centre cannot hold

—*W. B. Yeats*

Contents

Foreword

In December 2018, a young man sent me a short email that included this text:

> I'm a designer/technologist, and have been specifically working on the reduction of outrage propagation in social media. I've been developing a working group focused on ways of addressing it with colleagues in Silicon Valley and NYC. Many of my friends are early Twitter/FB employees who are also deeply concerned with recent trends.

He included a link to an essay he had published a few months previously, titled, rather hopefully: "Facebook's Problems Can Be Solved with Design." The article began with this simple image that the writer had created, which changed the nature of my thinking and research.

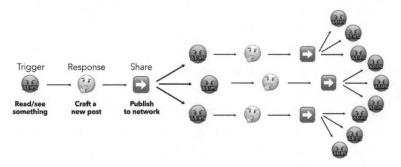

The writer, of course, was Tobias Rose-Stockwell, and you'll see this image again in chapter 9.

The image didn't say anything new; we all know about "virality" and the ways that social media can rapidly amplify anger and bad but emotionally evocative ideas. But what Tobias had done was capture a set of complex ideas and research findings about social dynamics into a single image that one can get, intuitively, in about two seconds, without even reading the few embedded words. This is one of Tobias's great strengths, and you'll benefit from that strength throughout this book. Tobias is a master of metaphor and intuition, which makes it easier and far more enjoyable to learn from him.

I am a social psychologist who has been studying morality and politics since the 1990s. I thought I had a pretty good grasp on things up through 2012. But by 2015, when I published an essay with my friend Greg Lukianoff, titled "The Coddling of the American Mind," I felt that there had been a sudden change in the basic operating system of society, flooding us with outrage, misinformation, fear, and just plain weirdness. It hit us first on college campuses, but it soon spread out to the rest of American society, and other societies, too. What the hell happened to us in the 2010s?

I continued to struggle with that question as things got weirder and weirder. The election of Donald Trump in 2016 did not cause the weirdness but amplified it and made researching it harder. How much of America's social and political insanity was caused by Trump's Twitter-based presidency and the intensely emotional reaction to him, and how much was a preexisting condition?

In early 2019 I was invited to write an essay on the weirdness for *The Atlantic* magazine. I accepted the invitation and began to outline my psychological arguments when I realized: I don't know enough about the technology to pull this off on my own. By that time Tobias had become an advisor to a nonprofit organization I had co-founded to promote constructive dialogue. He continued to impress me with his understanding of how social media interacted with human psychology in ways that could hinder or help authentic communication, so I invited him to join me as a co-author. By combining my knowledge of social and moral psychology with his deep understanding of technology and design, we produced an essay titled "The Dark Psychology of Social Networks: Why It Feels Like Everything Is Going Haywire." That essay laid out the story of how a few small design changes implemented between 2009 and 2012 led to the social and political chaos that began to engulf us all around 2014.

The essay was a preliminary statement of the charge that social media as it is now constituted is incompatible with the kind of liberal democracy and constitutional machinery that James Madison and other founding fathers had designed. We showed how social media inflamed "faction" or tribalism, and how it was now breaking society, civility, shared meaning, and democracy. This book, *Outrage Machine*, backs up that charge with a comprehensive presentation of the case, a review of the scientific evidence, and a wealth of stories to make the entire case vivid and unforgettable. We are now drowning in outrage. It wasn't always like this, it didn't have to be this way, and we can't go on like this.

This is a very important book. By the end of it you'll understand what the hell happened to us in the 2010s, and you'll see what we need to do, collectively and individually, to make it through the 2020s with our democracy and our sanity intact. We have to shut down the machine, or at least change its design.

Jonathan Haidt

The Time of Cruel Miracles

Imagine that one day a freak blast of cosmic rays hits the earth.

Spewed by a dying star millions of light years away, a barrage of gamma radiation blankets the planet and percolates through the atmosphere, exposing everyone on earth to a rapid mutation. In Marvel-movie fashion, this blast causes everyone on the planet to suddenly achieve superhuman status. Everyone gets the same new power.

As the sun rises that morning, all of humanity wakes up with the ability to snap their fingers and instantly make themselves invisible. The planet is still the same, plus one miraculous new human power. The day is punctuated by sheer amazement, awe, and wonder. Scientists and pundits spend their time marveling at this cosmic miracle. Introverts rejoice.

But by the evening, the world is in chaos. A looting spree overtakes the world's malls and banks. The worst actors, able to wander around and commit crimes with impunity, begin stealing whatever they can. Paranoia takes hold. People become afraid of what they cannot see. The National Guard is called in to protect businesses and government and much of the social order writ large begins to break down. The laws, rules, and institutions created to manage a visible society are suddenly rendered impotent as a small minority of awful, invisible actors wreak havoc on the system.

The chaos is contained only after an enormous collective effort to update how society should act with this new power. Within the turmoil, new rules for when invisibility can be used are put in place. Governments and companies begin deep research into thermal imaging systems, updating every security camera to try and account for invisible intruders. Phones are modified to be able to detect the unseen. Laws are passed to regulate when and where people can use their power. The new gift is frowned upon by many. It soon becomes morally unacceptable to use it in certain circumstances—a shameful thing to do if you're in a store or a church. A few religions ban it outright as sinful.

Yet invisibility becomes useful for those who don't wish to be beholden to the gaze of others. Some introverts begin rallying for a "right to be unseen." Certain stores, nightclubs, and cafés open only for the invisible. Some people begin whole relationships with others while never revealing their visible selves—citing the ability for their partners to discern their true nature without the material distraction of their bodies on display.

After a few years of chaos and turmoil, civilization adapts and begins to return to a peaceful state, with tight guardrails put in place to keep people from abusing their gifts. Collectively, humanity breathes a sigh of relief, remembering the first years of the miracle of invisibility as the "dark time." The fear slowly abates.

Then, a few years later, just as things seem to be returning to normal, it happens again.

Once again, a blast of radiation blankets the earth, and again all of humanity is given a new power.

This time every human on earth is now imbued with telepathy—the immediate capacity to beam one's thoughts to any and every other human in their own language. You can hear what people are saying *almost everywhere.*

This power is much more intimate, and at first, it's barely used. People use it for small thoughts updating their whereabouts, or modest requests for help. Others use it for sharing interesting news. But some activists realize they can capture nearly everyone's attention if they "say" things just the right way. If something is expressed with enough urgency and anger, it might gain the attention of the entire population.

Journalists soon find enormous value in this new superpower. They can report better on what's happening in the world—and it's soon a mandatory requirement of the job. Before long, politicians begin losing elections if they decide *not* to use their telepathy to ask for votes and criticize their opponents.

People realize their thoughts alone can make them celebrities. Ideas can make people famous, regardless of their accuracy. Rumors and falsehoods are sprayed with abandon. Emotions become instantly contagious. For most who choose to use telepathy, it becomes a regular addiction, something they can't imagine living without. Some are driven toward madness as they become ever more desperate for the approval of strangers. But many more are simply lost, overcome by falsehoods and confusion. Before long, there is an ever-expanding, overwhelming cacophony of opinions, commentary, and ideas.

Society again feels the stress, unprepared for the clamor of everyone's loud, explosive thoughts. The loudest, most vitriolic voices become those with the largest influence. False information outcompetes the truth. Confusion reigns. But no one can go back to the old ways. Many hate telepathy's influence, but collectively people are unwilling to give it up. It won't go away.

Studying this disruption, the astronomers look to the cosmos, and calculate a shocking truth: The dying star spewing these cosmic rays is actually *increasing* in speed. Every few years, with escalating frequency, all of humanity will now be imbued with new superpowers. How will society adapt to these rapid transformations? How will civilization endure?

Our ancestors would look at today's inventions, the internet and social media, and call them magic. They would marvel at our superhuman ability to spray our thoughts and feelings instantly anywhere on earth, coupled with the ability to communicate with impunity, invisible and hidden from view. A few of the many incredible gifts given to us with increasing frequency by our technology.

Today, as individuals, we're excited and eager to receive these gifts. We want the miracles to continue. We enjoy the benefits these new technologies bring. Yet with every iteration, every fresh power, the liabilities increase dramatically. These powers are given to us with increasing frequency, faster and faster. Without enough time, preparation, and foresight to adapt, society struggles to cope, until it does. But there is a deep darkness in the intervals between.

Introduction

This book is about a machine designed to make you angry.

The machine has a purpose: to inform you of what is happening in the world. Its job is to get you to pay attention, and it has become exceptionally good at this task. So good, in fact, that it has found out what makes you—specifically you—very mad.

In the process, it has divided you. It has fractured you into opinions. It has asked you to be for or against each new issue it has served.

This machine is our modern media system. It includes print media and digital media working in concert. It includes the device in your pocket. It includes newspapers and news feeds, algorithms and activists, politicians and pundits.

But the machine is not just made up of professionals. It also includes your neighbors, friends, and family members too. Without their knowledge, they have become participants in a continuous spectacle of news production and consumption. Many of them are so enmeshed in its gears that they cannot see it for what it is. Instead, they see outrages that *must* be shared, and threats that *must* be addressed.

When our minds fill up with threats, something strange happens. Instead of perspective, we see plots against us. Instead of people, we see enemies. We draw lines around our communities to defend the values we hold dear.

The new technology hardwiring us to this machine is social media. It has distilled, streamlined, and accelerated our participation in this spectacle of news better than anything that came before it.

This machine is a hard problem, because it also provides real value to us and society. It offers meaningful gifts of connection, ideas, and insight. But to use it, we must play by its rules. The rules are currently rigged against us.

This book is about how this machine came to be, and what we must do to adapt to its new intrusion into our lives. I know about this machine because I was there when it was modified. I know some of the people who helped change it. I know it was an accident—they didn't mean for it to turn out this way.

It doesn't need to be like this. We have a choice. Because we have been here before.

———————————

In these pages you'll find a story about the future, wrapped in a tale about the past. It is an exploration of what has been happening to society over one very strange decade: the 2010s.

Tracing the origin of our crisis of attention, I looked to the past for answers. I dug deep into the history and sociology of news, trying to understand the preceding eras of disruption brought about by new media technologies. I moved backward in time from the present day, to the beginning of the internet, to television, to radio, to the telegraph, to the postal service, to the printing press, and before.

On this journey I found something very surprising: Despite how fractured and dystopian this moment feels, it isn't fundamentally new. It is instead the most recent phase in a lengthy cycle of disruption and renegotiation spanning centuries. Yes, the technologies built to capture human attention at scale are more powerful today than they have ever been. Yes, their influence feels more threatening than ever. But their invention follows a historical path that has been well traveled by our species. This path follows a process of increased amplification of viral ideas and speech, accompanied by an explosion of unintended consequences, then followed by controls and guidelines placed to try to maintain a kind of epistemic balance—a balance of good information over bad. As groups of people begin observing the harms of these new tools, they push to renegotiate their implicit contract with them. They learn to demand more from their media. We are in the midst of one of these cycles now, a period that feels like we are trapped in a dark valley.

Each of these periods of major disruption are punctuated by confusion, disorder, and violence as small groups of activists and innovators exploit these tools to advance an agenda. Adaptation eventually comes, but at a cost to some small fraction of amplified speech. Not all speech is curtailed, but the speech that exploits our natural tendency to engage with the extreme, the inaccurate, and the inflammatory. Society recovers as it figures out how to establish a semblance of accountability, tolerance, and cohesion based on who has access to the most minds. To those of us born into the age of modern journalism, this most recent disruption feels like a slide into chaos. But there is a pathway out.

This book focuses on social media, the newest tool to initiate this cycle. Throughout its pages, you'll find research, diagrams, principles, and sketches of solutions to the problems we're facing. These resources are cataloged in a living appendix which will be kept updated at http://outragemachine.org. The book is divided into five parts, each exploring different pieces of our relationship with it:

Part I, Making the Machine, sets up the major components of the modern viral internet—smartphones, social media, and news—and how they work together to command our attention often against our better judgment. It follows the story of how I found myself at a strange intersection of online activism and the earliest social media.

Part II, Powering the Machine, defines the elements that keep it in motion—algorithms, metrics, and moral emotions—and how they combine to create cascades of outrage and cancellation that have come to define our culture today. This system is now dominated by groups looking to advance agendas, from activists, to politicians, to conflict entrepreneurs. We'll also explore how every new technology goes through a period that I call a *dark valley*, when the harms caused by tools are obscured by their mass-adoption.

Part III, History of the Machine, goes into detail about the original attentional process that makes the machine work—our desire for news. Here we'll learn how journalism began as a salacious system of mass-attention capture, maturing into an organ of semi-objective truth-finding that all democracies need to work.

Part IV, The Cogs in the Machine, focuses on the three main gears that need to be kept in good working order for the machine to run smoothly: truth, trust, and freedom of speech.

Part V, Rewiring the Machine, explains what's at stake for democracies if we don't make tactical adjustments. I end with sketches of solutions, and ways that we might start designing these tools to serve us, rather than the other way around.

Let's begin.

OUTRAGE MACHINE

PART I

Making the Machine

Chapter 1

Empathy Machine

It's strange to start a story about technology in a rice field in Cambodia, but that's where it began for me.

I was sitting under the shade of a big thorny bush looking at a sad little stream splitting two massive mounds of earth in half. Water buffalo roamed in the fields below me, rummaging among dry rice stalks. It was over a hundred degrees, just past noon, and I had been up since four a.m.

The mounds formed a squat embankment that stretched into the distance in both directions away from this lonely little stream, which pooled beneath us, marking where an ancient irrigation system once stood. A young Cambodian monk—whom I'd met just the day before—stood next to me, wrapped in a loose saffron robe. He spoke in short, emphatic bursts of broken English. "You see? The dam. It broke!"

If I squinted, I could almost see it. At one point, these two piles of earth connected, and blocked this little stream, causing the water to pool. Judging by how overgrown this thing was, that point was a very long time ago.

He had brought me here, to this sad spot under the sticky Southeast Asian sun with an agenda: He wanted me to fix this broken dam. For some reason, he thought I could do it.

I certainly didn't.

I was not an engineer, a hydrologist, or a trust-fund kid. But this monk wanted me to rebuild a reservoir and was looking for $15,000 to do it. I barely had enough money to finish my backpacking trip.

What began that day was a very strange chapter of my life.

Nearly seven years later, I would stand in that exact spot and watch the gates close on a massive concrete spillway I had built with the help of dozens of engineers, hundreds of volunteers, thousands of donors, and *hundreds* of thousands of dollars I had raised. As the water slowly pooled into what would become an enormous irrigation system, providing water to over five thousand poor rice farmers in the middle of nowhere, a single thought bounced around my head:

Thank you, internet.

Let me back up for a moment.

Today, it's particularly difficult to speak about a subject without first addressing one's personal history. So let me start by taking a moment to tell you a bit more about where I came from. My name is Tobias. I grew up in a suburb of the Bay Area in northern California in a modest middle-class home. My father is a journalist, programmer, and musician, and my mother is a librarian who became a college professor. My family's background, according to 23andMe, is "broadly northern, western, and southern European," with a dash of Japanese that somehow weaved its way into my distant family tree. I come across as a pretty standard white guy. My nephews are Black, my nieces are ethnically Jewish, my partner is Chinese American, and our children and their cousins will be unmistakably intermixed. My family's genetic lineage is a converging reflection that might best be simply called American.

I struggled in school and was diagnosed with ADHD as an adult. I made it through my degree studying art and psychology by doing some subversively creative projects for my senior thesis, spreading viral memes across campus in an era before most people even knew what a meme was.

When I turned twenty-two, I worked for a year to save money for a solo backpacking trip through Southeast Asia. During this trip, I was compelled to immerse myself in new cultures as a tourist exploring beaches and hidden villages, but also spent almost half my trip volunteering, including doing administrative work at a home for HIV-positive kids in Thailand. Toward the end of my travels, I found myself in Cambodia exploring the temples of Angkor Wat. Over a thousand years old, they're an iconic jungle temple complex—and the largest religious monument in the world.

After a week of exploration, I returned one evening to my guest house, a small, creaky old place with wooden rooms that cost two dollars per night. The door of my room had some commonsense guesthouse rules: "We are not responsible for gold and gems left in your room." And, "Please keep your guns and explosives with the management, thank you."

As I wandered out for dinner (seventy-five cents for the best rice and chicken of your life), I found a saffron-robed monk sitting at a table speaking to one of the other guests. He was a small man who spoke so fast in broken English that it was a struggle to keep up. Curious, I joined his conversation, and before I knew it, he had invited me to visit his village in the countryside. It was a simple offer that resembled local kindness: *Come to my home, meet my relatives!* After a moment of hesitation, I agreed to go. It ended up being one of the most consequential decisions of my life.

The next morning, at five a.m., a parade of six motorbikes showed up on the dirt road in front of the guesthouse. Four were occupied by orange-robed monks and their chauffeurs. Two other motorbikes had drivers, and one of their tiny back seats was reserved for me.

As we drove on pockmarked sand roads into the dry countryside away from the touristy city, we moved a step back in time. Rice farmers resided in tiny houses built on sticks, and water buffalo plowed the fields. We quickly became the attraction as we putt-putted through the rutted roads.

After several hours, we reached a large, dusty wooden pagoda set back among coconut palms. The monks sat me down on a bamboo mat on the tiled floor, and instead of meeting their families, I found myself surrounded by dozens of locals: the village elders, the commune council, and a huge group of farmers. They began a presentation. One by one, each village elder stood up and said a permutation of the following: "Thank you for coming! We have been waiting for you. We appreciate that you've agreed to help us rebuild our reservoir."

What reservoir? This all came as a surprise for a number of reasons. The monk hadn't mentioned any reservoir in our initial conversation, just a community visit. He also didn't tell me that there would be so many other people from the surrounding villages anticipating our arrival with bated breath.

After the confusing presentation, I sat there quietly, looking at them wide eyed. I tried to gently explain that they had the wrong person. But that message was either intentionally ignored or lost in translation. They smiled, fed me coconuts, and forced me to lie down and take a nap.

When I awoke in the early afternoon, they smiled more, put me back on a bike, and drove me out to that sad-looking hillside under the baking Cambodian sun where a tiny river snaked through a mound of old earth. It didn't appear to be anything close to a reservoir. It looked like an overgrown pile of dirt, rising above a sprawling field of dusty rice stalks.

They began to explain. The average member of this community lived on two dollars per day, they said. The reservoir could double or triple the income of the community living there: all five thousand of them, spread across several villages. They estimated it would cost about $15,000 to purchase enough rice so that they could feed the community while they rebuilt the earthen embankment, dug out the old canals, and flooded the rice fields. This was the real ask—the real reason the monk had brought us here.

Ahhh, I thought. *I'm playing the role of the wealthy foreigner who they think can pay for their project! Yeesh.* Feeling skepticism rising in my throat, I again tried to explain the confusion, and that they must have the wrong guy. I was a young backpacker living on rice and noodles. This was not something I knew how to do.

The monks smiled and happily kept ignoring my attempt to clear up the confusion.

After the broken reservoir, the monks led me to a trail nearby and shared a local secret—a beautiful ancient temple sitting on the hill above the pagoda. Walking among the ruins of this thousand-year-old sandstone structure, they spoke about how they grew up in this community and how they'd narrowly escaped being killed during the fighting of the civil war that had happened there. They showed me a bomb crater that one of the monks had hidden in as a child to save his life during some especially intense fighting. They showed me where the remnants of an antiaircraft gun had been placed adjacent to an old schoolhouse. They pointed to a spot where one of their siblings had lost a leg to an American-made land mine.

As the sun lowered and we began our long journey back, I reflected on what had happened to that country, and the world these young monks had grown up in. Civil war had come to this place and left its mark everywhere. The number of missing limbs among the villagers, the water-filled blast craters, and the bullet-hole pockmarks in ancient walls spoke to the struggle that happened here. Suffice it to say, these people had endured a lot.

As the parade of saffron-robed monks dropped me off back at the guest-house, something stirred in my chest. I was feeling a pang of empathy and sadness observing the plight of these villagers. Touched, I told them: "OK. Let's meet back here tomorrow and talk some more."

I decided to stay and help. Over the next few weeks, feeling connected to the monks' desire to rebuild their community and fascinated by the history of civil conflict in this place, I dug in. I spent time with the monks, trying to help figure out how to raise capital for the project while learning all I could about the history of the area, the tradition, and the decline of this once-thriving country. I did dozens of interviews with local members of the community and heard their stories.

What I learned was unquestionably horrific.

Cambodia suffered through one of the most violent and deadly periods of the last century. The year was 1969, and the Vietnam War was in full force, supported by widespread panic about the expansion of communism. The Nixon administration intensified a secret bombing campaign on the eastern portion of Cambodia in a vain attempt to hobble Vietnamese troops hiding there. The results were disastrous for civilians.[1]

A small group of radical communist idealogues called the Khmer Rouge used the chaos to their advantage. Over several years, they steadily captured territory, and in 1975, after years of fighting through the countryside, they captured the capital city of Phnom Penh and toppled the US-backed government.[2]

Within a year, they had established what one academic called a *political religion*—a combination of religious and political doctrine. It was radical, extreme, and ultimately genocidal.[3]

Those who challenged the orthodoxy of the party weren't simply ostracized or condemned. They were often taken away and murdered. Through forced rituals of public humiliation and brutal practices (like killing and defrocking every monk in the country), the Khmer Rouge made a cult out of a political movement. It became an ideological, rather than ethnic, genocide: Believe or be killed.[4]

In just ten years, this gentle land descended into chaos, brutality, and slaughter. In the period between 1975 and 1979, Cambodia saw the largest per capita loss of life in any nation in modern times.[5] Scholars estimate that more than a million Cambodians were systematically starved or murdered by self-inflicted political extremism, eliminating anyone the regime suspected to be sympathizers with the outside.[6]

Neighboring Vietnam invaded in 1979 and deposed the Khmer Rouge, thrusting Cambodia into another long and violent period of guerilla warfare. Until the late 1990s, Cambodia was still an extremely dangerous destination, with regular bombings, kidnappings, and random outbreaks of violence.[7]

Every Cambodian I spoke with above the age of thirty had a horrific story of survival, losing a loved one, or narrowly avoiding death. Listening to these stories, I was touched, frustrated, and outraged. A moral fire had come alive in my chest. These earnest monks were trying their best to rebuild after a terrible tragedy. They were quite honestly asking for help. The best I could do was try.

So, I turned to the internet. I drafted an email to my friends and family about my experience meeting them. It included an impassioned reference to the difficulties they'd endured and a focus on the suffering they had faced. I had a mailing list—one that I'd used to send out regular updates about my journey through Asia in long missives that read like journal entries about my adventures. A simple list of all my closest friends and acquaintances.

When I clicked Send on that email, something extraordinary happened. Friends forwarded it to their friends, and their friends, and so on. It caught fire online, becoming a small viral sensation among my extended network. It did so because of one remarkable connection.

One of my email recipients was a larger list-serve, sitting inside a website that a programmer friend back home had built. This server had about two hundred friends-of-friends on it, linked together with simple profile pictures, bios, and a common forum fed by our emails. The list included mostly young twenty-somethings (like me), who were connected to the underground music scene back in California. This site, hand-built to serve our community, was a bespoke piece of technology that barely had a name: it was a social network.

This online community blasted my email out well beyond its natural range, into the mailboxes of a vast number of third-degree connections. Before I knew it, I found a broad swath of interest in helping this unknown group of Cambodian villagers. Friends and strangers alike would email me with interest, asking in numerous ways: *This is amazing. How can I help?*

And quite suddenly, there was support. People wanted to be involved. It soon seemed plausible that I could actually *do* something for this group of monks. Using a powerful story and a simple social network, I'd manifested a viral cascade of compassionate interest for struggling, rural farmers living at the edge of nowhere. Something had clicked.

At that moment, I was told by the monks that it would take roughly two months to help them complete the reservoir. *OK*, I thought. *Two months. And $15,000. That is basically the cost of a car. I can do this.* I flew back to the States, and back to my job building websites. I saved up a little more money for my travel expenses, and after a few months boarded a plane back to Cambodia.

Upon my return to the country, I was dealt a slow and steady stream of blows to my optimism. What I learned was that the project was *huge*, far larger than I thought it would be both in terms of physical size and effort required. The reservoir was more like a lake; the embankment alone was half a kilometer long, a fact obscured by the dilapidated nature of the weathered old structure. It would require a huge concrete spillway for water management, with engineering schematics, hydrological plans, and detailed mapping— a sizable infrastructure project for any local municipality. I also learned the whole area had land mines and unexploded ordnance scattered throughout the countryside.

Those two months, through an ever-increasing escalation of commitment, fascination, setbacks, and persistence, turned into nearly seven years of my life in Cambodia.

Through that initial viral cascade of attention to the project, I was able to leverage support from friends and strangers alike. I bought a domain, built a website, and called the effort the Human Translation Project, because I knew this new capacity for connection using the internet could increase people's compassionate ties to previously hidden causes. This connectivity, I felt, transcended language and traditional geographical barriers.

I incorporated a nonprofit entity, recruited a board to help run local fundraisers, and collected hundreds of small donations to implement the project.

Before I knew it, I was running a small organization with a dozen employees and volunteers working to help the area rebuild.

We recruited a number of experts to work on the reservoir, namely a volunteer group called Engineers Without Borders, to help design the spillway, the embankment, and manage construction while I worked with local staff. We partnered with another organization to remove unexploded land mines, and cleared dozens of acres where farmers had regularly been losing limbs after accidentally triggering them planting rice.

Over the years, that poorly scoped $15,000 estimate from the monks ballooned to more than $250,000 in construction costs. After a steady stream of successful fundraisers and a flurry of grant writing, we secured enough money to start construction. By 2009, and after years of setbacks, profound complexity, and arduous work, the reservoir was completed. As I watched the reservoir fill for the first time, I felt a deep and abiding sense of potential. I knew a project like this could be duplicated. I knew more people deserved this kind of support. I knew more people could get involved with projects like this *if they just knew they existed.* This kind of viral attention could change lives.

Ready for a new challenge after nearly seven years, I wrapped up my involvement, handed the reins over to local staff and community, and transitioned out of the country to return to the Bay Area to figure out how to scale up efforts like this. I carried a few key lessons with me as I went:

The first was the recognition that a viral cascade of connectivity, leveraged by the internet and a simple social network, had given me power and resources I hadn't previously had, to solve a problem that otherwise wouldn't have been visible. In short, the internet was a powerful force for raising awareness and increasing emotional attention to unknown causes. There was an opportunity here to do more of this.

The second was that moral extremes often have terrible outcomes for societies. A radical political ideology had wrecked Cambodia, and it was taking generations to pick up the pieces. This experience informed my worldview about the potential of peaceful societies to quickly go from tolerance to tragedy when their politics turn to extremism and violence.

Last, I lost faith in the ability of good intentions alone to solve problems.

I had begun to feel the first stirrings of skepticism towards my own righteousness. None of the simple narratives about the project had turned out the way I expected. Almost every single expectation I had about how to help turned out to be wrong: My idealized assumption of the monks' capacity to execute on the project's needs. The cost overruns. The land mines. None of it was easy. The reservoir was a hugely complex task, and my righteousness could take me only so far before it met the hard earth of reality.

I certainly didn't know that within a decade these three lessons would converge into a strange dystopian vision of the world—one we're becoming increasingly familiar with today.

From Empathy to Outrage

In 2009, I wrapped up my projects in Asia and moved home to the San Francisco Bay Area, where I had grown up. Back in the United States, the social web was just coming online. Friendster, Myspace, then Facebook and Twitter had begun their rapid, tendrilous expansion into the personal lives of Americans. Even living overseas, it was hard to ignore their emerging ability to capture our collective attention. Within the first few months of being back in the States, I was invited as a guest of honor to a luncheon in San Francisco co-hosted by one of my donors. The luncheon was memorable because it was co-hosted by His Holiness the 14th Dalai Lama who was there to bestow awards for humanitarian work on a select group of social entrepreneurs and nonprofit workers. While I was one of the award recipients, I was far from the most notable person there. Among the many distinguished guests, at my table, sitting next to me, I found a young man named Cox with kind eyes who had been invited as an attendee. He would soon become one of the most quietly influential people in the world.

Chris Cox (known by most around him just as Cox) was Facebook's VP of Product, and the man responsible for many of the company's critical design decisions. We found ourselves in a long conversation about the transformative power of social media. I shared my story about the viral beginnings of my work in Cambodia, which he found thrilling. He espoused a vision of leveraging social media for social good, and the inherent power of connectivity online. We became friends on Facebook, and a few weeks later I would travel to Palo Alto to visit Facebook's early offices and discuss the emerging potential of these tools.

In the months following, fascinated by this promise of activism amplified by social media, I linked up with a cohort of young social entrepreneurs, documentarians, and programmers who also saw this technology's tremendous capacity to connect humans in the most positive ways. For those of us interested in fixing humanitarian issues at scale, these platforms looked like a gift. Our assumption was simple: Global problems would be solved if people knew about global suffering. If we could make the invisible plights of those in need visible, we thought, we could imbue others with the sort of righteous empathy we felt every day.

And in many ways, it worked. In the following years, entrepreneurs and activists in my cohort were responsible for campaigns that attracted billions of page views and pioneered a new type of advocacy online. We built projects that catalyzed moral action at scale. Several friends made a short film called *Kony 2012* that, at the time, became the most viral piece of content in human history—that is, before exploding spectacularly in a cloud of conspiracy and online scorn.[8] My friend Ben began a platform called Change.org that helped people sign petitions online, becoming one of the largest social activism sites on the planet, helping to mobilize thousands of movements locally and globally. Another friend in Chicago ran the technology infrastructure that propelled President Barack Obama to two terms in the White House.

We built our efforts on the back of a series of digital products that exponentially increased our reach: the news feed and the Like button at Facebook; the launch of the Retweet button on Twitter; and a ranking algorithm that began to increase the reach of emotionally charged content. With these specific new features—most of them created with positive intentions—these efforts became more viral than ever.

As these various projects dominated headlines, galvanizing philanthropic and political action, we learned much about what worked and what didn't when it came to motivating people en masse. We became more and more interested in the psychology underpinning empathetic responses, the things we could do to trigger people's emotive attention using social media.

Like many social entrepreneurs at the time, we were obsessed with the idea of empathy as a tool to change human behavior. I was convinced that any increase in the quantity of empathy would make the world better, that empathy was literally one of the most important feelings humans could feel. A popular refrain of that era was Obama's speech on the so-called empathy deficit, and how, in itself, empathy is "a quality of character that can change the world."[9]

Social media was the most important new tool we had for triggering empathy. Though it's hard to see today, the optimism around these tools was palpable and infectious in the first years of their use. We assumed that this technology was virtuous in itself. A notable example of this was during the sustained protests of the Arab Spring, when we watched the real-time upending of brutal authoritarian regimes, and we, along with most of the Western world, praised the roles that Facebook and Twitter played in facilitating them.[10]

Social media appeared to be a liberator, a one-way street toward more liberty. More democracy. More visibility. More information. A popular refrain throughout Silicon Valley during this era could be summarized as: "Don't overthink the purpose of products you build. Mark Zuckerberg created Facebook to get girls, and now it's toppling dictators in the Middle East." The subtext here was that these tools had their own kind of hidden agenda—one that was inherently beneficial for humanity—regardless of intent.

And while we were right about the significance of social media's impact, we were very, very wrong about its inherent goodness.

Righteousness Is Not Always Right

Yale psychologist Paul Bloom has argued that one of the greatest misconceptions about empathy is that it is a benevolent emotional feature above all else. Instead, he draws a clear line between empathy and what he calls "rational compassion," noting how many problems can come from feeling empathy. The empathetic impulse to feel another's pain is very helpful in interpersonal relationships with lovers and friends, he says, but as a guide for moral decision-making, it is flawed, parochial, and inconsistent.[11]

It's also frequently manipulated for political gain. Bloom argues that empathy has historically played a role in the lead-up to every major war, supported by a concerted political effort to outline the specific dehumanizing acts of the enemy. In democracies around the world, this is done by describing the atrocities of the opposing regime. The victimization of Yazidis by the Islamic State brought the US into Syria. Highlighting the civilians massacred by Saddam Hussein justified the drive to invade Iraq. The South Vietnamese casualties of Ho Chi Minh were used to validate the Vietnam War. Empathy always plays a distinct role in propaganda, explicitly asking the viewers to feel the suffering of the enemy's victims and calling on them to respond with condemnation and moral outrage.[12]

Many in my cohort understood this power implicitly, even if we didn't recognize these tactics could also be used for ill. We created content for causes that mattered to us the most: ending poverty, bringing war criminals to justice, feeding the hungry, supporting brilliant politicians we believed in.

We knew how to provoke others to care using these new tools and how to call out those who weren't doing enough. We learned how to elicit strong emotions, to tug at heartstrings and make people feel something. Social media was there to help us share these urgent problems with the world, because these problems *needed* to be shared.

What we didn't know was that we were just at the beginning of a very new, very disruptive trend. As a small group of technologists, activists, designers, and filmmakers, we were on the advancing edge of a revolution influencing human behavior at scale.

As we worked to find the triggers and stimuli to create these empathetic responses in others, we were, along with the rest of our cohort, opening a sort of Pandora's box. We had found a control panel for emotive action, a guaranteed way to extract attention from others. A way to get people to feel something for those they had never met. And these tools would soon be available to anyone.

An example of this can be seen in the massive initial growth of *Upworthy*, a news site dedicated to sharing positive stories about the human condition started by Eli Pariser and backed by Chris Hughes, one of the cofounders of Facebook. Its success began with an innovative process of aggressively testing dozens of headlines to find a viral "hit": something that people would see, click, and share rapidly online. During *Upworthy*'s prime in 2012, these tactics made it one of the most popular news sites in the world, with tens of millions of views per article.[13] But others soon took notice. With no barrier to entry, this innovation was quickly copied, and within a year, these kinds of viral headlines were everywhere, adopted by every digital news site looking for a traffic fix. This well-intentioned innovation was the inception of what would come to be known as "clickbait"—the viral scourge that began a race to the bottom of the internet.

By the middle of the decade, these data-driven tactics of extracting emotional engagement had made their way into the mainstream. A new economy of attention had emerged, playing on empathy and outrage, shaping the way we discovered and interacted with information. Many online news organizations were forced to change their journalistic approaches to stay relevant. They were rigorously analyzing audiences for clicks, using hyperbolic headline optimization, and auto-playing emotive clips to suck users in. Moral language began filtering into straight news coverage. At the same time, propaganda from anyone, by anyone, began seeping into our news feeds. Emotional manipulation for profit had become a commonplace practice, available to all, promoted by algorithms—and just in time for the 2016 presidential election.

That year, for the first time that I could recall, political animosity in America began to resemble the fractured extremes of a nation on the edge of civil conflict, not far off from where I began my career. A level of disdain, fear, and moral panic had entered the cultural zeitgeist. I could clearly see the machinery that got us there: a system inadvertently designed to make people outraged. Purity tests. Cancellations. Partisan animosity and manufactured anger. Outright hatred. Fear everywhere. All of it toxic to the foundations of a democratic society.

What my friends and I had failed to understand so many years ago was a strange truth: If you can find the control panel for empathy, you can absolutely push the buttons of outrage. And once this machinery was out in the open, it couldn't be put back in the box.

IN SUM

We began this chapter exploring how I found myself at a strange intersection of civic activism and the internet. Using a simple social network, it was possible for me to generate a viral cascade of attention focused on an obscure but meaningful cause I was passionate about.

When I arrived in Silicon Valley in the years thereafter, I joined an ambitious new effort to use this new tool—social media—to give anyone the capacity to amplify any cause they cared about. The underlying assumption behind these efforts was that connectivity increased feelings of empathy, and empathy was inherently good for humanity. Empathy, however, as Paul Bloom has shown, is a double-edged instrument that can just as easily be used to evoke outrage.

The creation of these tools coincided directly with something else that was happening in society as we all began to come online: Our emotions were becoming more contagious than ever before. This maps with a remarkable and accelerating trend that we will learn about next—one that is fundamentally changing our species. To understand how we got here, we'll need to step further into the past.

Chapter 2

The Feed

Let's go on a journey through time.

It's 16,000 BCE.

You're an unremarkable human being living on earth. Your floor is made of dirt. You live in a tribe—your extended family—that moves with the seasons. You were born into a life that is in sync with the natural world. You are aware of fickle and capricious gods that rule the landscape around you. You have been told how to assuage them, learning the rituals and the rites. Your parents and the elders of your clan hold the knowledge—all of it—in their heads. They know what is true. They are the stewards of the past and the future. You trust them with what is real and what is not, and what is right and what is wrong. They learned it from their elders, and those elders learned from theirs, and so on. New knowledge rarely enters your world, and your prospects for finding more of it are grim. If it does come, carried by an outsider, it's suspect and dangerous. Your feed is your self-contained tribe.

It's 2000 BCE.

You are a farmer. You live in one place and work the land with a community, growing things with purpose and intention. Your years are cyclical and repetitive. When you trade crops, you sometimes hear news. This new information is spread with the goods you exchange. The message, brought by travelers, is from long ago in a distant place. The stories are fantastical, sometimes terrifying, but always interesting. You listen to these tales about a world you will never experience. They open your mind and your imagination, even if you will never leave.

It's 700 CE.

You live in a town. You are poor, but most everyone else you know is, too. You trade your services for goods, for food, for the things you need. The town has a rigid hierarchy, and everyone knows where they stand. Those at the top have a power: They can read lines, etched on soft material, that carry meaning. These lines are written by others but separated from the creator. Knowledge living on its own—a box containing words—a book. People who can read tell you what it says, and it is clear: Things are how they should be, and their authority should not be questioned.

It's 1500 CE.

Papers arrive with events written on them. They are sold or read aloud in public, and beckon you to participate in the dramas of the wider world. Asking for your attention and concern, they serve opinions in the form of little boxes of text. These opinions light your brain on fire, and suddenly you carry opinions, too, about what *should* and *shouldn't* happen in the world. About how to do new things—ideas and concepts you had never considered.

It's 1800 CE.

You live in a city, which is part of a thing called a nation. Papers are everywhere. You can read them. Your nation is a huge place full of ideas. Somehow you can make some sense of what is happening in the whole world, you think. It's all very complicated, but it's not a mystery. If you want to understand what is happening, you just need to pay a few coins, and you are given a paper with vast knowledge from the recent past. You learn something new every time you pick one up.

It's 1900.

If you live in the right place, you can go to sit in a dark room with other people, where you watch things: a giant rectangular box on the wall. Moving pictures dance inside. There might be images of news. Glimpses of wars. Silent shows that are projected on the screen. These captured images make you feel a vast spectrum of emotions. They're mesmerizing.

It's 1930.

A new box comes into your home: a radio. It enters your living room, speaks to you, plays you music. It tells you things you hadn't heard before, using a real human voice. You are aware that this is happening now, for everyone, all at once. It feels for the first time that someone from far away is speaking to you in your own home. It asks you to be on its schedule, and so you sit down and listen to it. It plays you music and narrates stories to you. It reads news faster than the newspaper can be delivered. You now have access to a real-time feed, a curated stream of information. You feel like you're a part of something much bigger, a shared experience that everyone who has one of these boxes also feels. Suddenly we are all connected in time, listening together.

It's 1950.

A new box arrives in your home, just like the radio but with pictures. It pushes images at you with sound: music, ideas, moving all at once. Dazzling showmen in a tube. It feels significant, powerful. We're all now watching a stage at the same time breathlessly. A performance in a box. It includes images of our leaders speaking for the first time. News becomes entertainment. Entertainment becomes news. People who look trustworthy are telling you to buy things, to fight, and to vote. So you do. Why would they lie?

It's 1980.

If you're lucky, you have access to a newer box. This box has a keyboard. It lets you put things into it and gives things in return. At first, it's like a TV displaying a book. You press buttons and it gives you back words. It shuffles inputs, asks you to think. But it thinks a little bit on its own, too. It gives back images, pixels that move. Games you can play. It interacts with you. It computes for you. You ping, it pongs. You type, it returns.

It's 1995.

This box is quickly replaced by a better one. Then a better one. Then a better one. It shrinks every time. These boxes are now connected. They talk to one another. Now your box speaks to everyone else's box. In real time, you can see people speaking back to you. A community made of words and pixels.

They are normal people, not showmen. Everyone has a place to be. There is a page for everything. The boxes begin to show up everywhere. In offices, libraries, schools.

It's 2007.

The box shrinks so small that it fits into your pocket. When you have a free moment, it will tell you something fascinating. Something you didn't know before. The box is full of ideas and opportunities at all times. If you're bored, you don't need to be. You can put your hand in your pocket and find a thing to occupy your brain. The space for you and your own thoughts shrinks, because everyone else's thoughts are right there with you, constantly.

It's 2009.

The tiny box starts asking you for your opinions. It buzzes and pulls you in, telling you what other people think of you. What they like. What they shared. The news of friends, family, relatives, and strangers is now mixed in with papers, journalists, pundits. You cannot tell them apart. You tell it what you're doing, and other people chime in. It pulls you in even more.

It's 2012.

The box starts telling you things that make you very mad. You notice you're inside the box longer and longer. It lights your brain on fire with anger. You cannot put it down. What *should be* and what *shouldn't be* are ever more present in your mind. You turn off the buzzing. But it doesn't keep you away. You've been trained. You now look whenever you're bored, or sad, or whenever you feel a nagging sense of self. Whenever your mind has space for reflection, it wants to go back to the box, to be filled by other people's opinions. To have opinions about those opinions. To feel connected to the feed.

It's 2023.

The boxes are the first thing we see when we wake up, the last thing we see when we go to sleep. The boxes beckon for us to contribute. They ask for our offering, promising money, notoriety, respect, fame. They become harder and harder to ignore. Before we know it, we live more inside the boxes than outside. The box has become the world.[1]

Chapter 3

The Overwhelming Present

> What information consumes is rather obvious: it consumes the atten-
> tion of its recipients. Hence a wealth of information creates a poverty
> of attention, and a need to allocate that attention efficiently among
> the overabundance of information sources that might consume it.
> —*Herbert Simon,* Designing Organization
> for an Information Rich World

We were born into an information revolution.

We consume exponentially more content on a daily basis than our ancestors did. Every generation increases the quantity of information produced
and consumed, and that quantity tracks on a steep curve: Our parents consume more content than their parents, and so on. We are living through an
ever-increasing explosion of the availability of information, and the velocity
of its injection into our minds.

But our brains didn't evolve to process this much new data. All information can be measured in bits—a binary digit, a one or a zero. Any stream of
characters, language, sound, or images can be distilled down to this basic
unit. It's the code used by computers, made legible to us in the words we see
presented on our screens.

Our conscious minds have a hard speed limit—a bandwidth restriction
that we cannot get around. We can process a maximum of 126 bits per second of language. A single conversation with another person takes around
sixty bits per second of our mental capacity—half our average attentional
bandwidth. According to the psychologist Mihaly Csikszentmihalyi, this is

why processing two conversations at once is difficult, and tracking three conversations simultaneously is nearly impossible.[1]

A 2011 estimate by the neuroscientist Daniel Levitin found that the average American consumes five times more information than they did in 1985—roughly 242 billion bits *a day* in their leisure time alone. That's the equivalent of reading 175 newspapers daily.[2]

But while we are often exposed to more than 120 bits per second, our brains have a unique tool that allows for us to tune out the noise: an attentional filter. This filter is with us during every waking moment, screening out the excess inputs. It evolved to help our ancestors decide what to focus on when. Without it they would have been lost.[3]

Every species has a different type of attentional filter. Cats track rapid movement of small objects and the high-pitched noises of potential prey. Dogs use their filter to discern between a huge number of potential scents. But humans are unique. Our filters are hypersocial. These filters evolved as a result of living in a collective, causing us to naturally focus on social interactions. We innately want to know what the other people around us are doing: Which actions are looked at as right and wrong? Who is important? What is our standing in the community? As a result, our attentional filter is hyperattuned to our social worlds.[4]

Today our brains spend much of the time plugged into a network saturated with social information. We find ourselves obsessed with metrics that reflect these social signals. Despite our best efforts, it feels nearly impossible to pull away. How did we get here?

The Social History of the Internet

The revolution began, as all revolutions do, with idealism and great optimism. For the earliest internet pioneers, many believed it was the beginning of a utopia.

This first internet age began in earnest in the 1980s and 1990s as research into microprocessors allowed for the rapid size reduction of supercomputers. Gradually, these machines made their way into our offices and homes.

There were real barriers to entry: computers were still expensive, and if you wanted to get online, you had to connect to an ISP with a complicated modem. You also needed knowledge of network protocols. And if you wanted to really build things, you needed to read and write code. For these and other reasons, the early internet was filled with a homogenous group of humans who tended to be progressive and highly educated with an inclination toward counterculture ideals. Many of the earliest pioneers worked at universities and research institutions, where the US and Europe had made major investments in new technology during the Cold War.

In 1989, at one of these agencies, the European Council of Nuclear Research (known as CERN), a young English programmer was struggling. He was trying to share scientific documents across the mess of different networks that had been developed to exchange data across institutions—a tedious process full of friction. Tim Berners-Lee saw that these networks were missing something—a layer that could tie these systems together with a common interface. He coded a new set of tools: Hypertext Markup Language, Hypertext Transfer Protocol, and Universal Resource Locators (HTML, HTTP, and URLs, respectively), creating a uniform way for these disparate protocols to connect and display their information visually. These tools made it easy for anyone to see which documents were presented on a server with text and images. Berners-Lee called these tools a documentation system, but they were far more than that. They became a new standard method for people to see online content: as a website. Perhaps understanding the awesome ambition of what he was building, he called the address by the abbreviation "www," short for World Wide Web.[5]

Within a few years, his standard exploded, and thousands of new websites opened up access to anyone with a browser and an internet connection. The language for display, HTML, was simple and straightforward to code in. Anyone with a server could build and host a web page, and as long as someone knew your address, they could read what you posted.

In this early era, new publishing platforms like Geocities and Blogger emerged, which enabled anyone to publish their content online without a server, and without the critical eye of a journalist or editor. Publishing was now a democratized, frictionless, and cost-free endeavor. It was egalitarian, open, and free—paid for using an old-school business model: placing ads next to your content.

There has been some speculation that this business model could have been different, and that better revenue streams like subscriptions or micropayments could have been embedded into the design of the earliest web. Many of the attentional problems we face today are the result of what the scholar Ethan Zuckerman calls the "original sin of the internet," an ad-driven model of attention-extraction underpinning most web services.[6]

The reality is that advertising dramatically outcompeted alternative business models for a variety of reasons. There was no payment solution available in the early web (credit card companies had no realistic way of integrating online payments for years, PayPal wouldn't be launched until 2000). Also, the infrastructure of the early internet was shockingly expensive to build, and investors willing to throw money at it sought a high return on their risky bets.

But the biggest reason for an ad-driven internet ended up being psychological: something free is preferable to something you pay for. Users were more willing to give away some fraction of their attention for free content. As new users flocked to publish on these early ad-driven platforms, very few websites actually took off. Most pages were simply not discoverable and languished in obscurity. This began to change when a few early search companies introduced "crawlers," tiny bits of code that would scan websites to index them, and put them into a searchable database.

Of the dozens of early search services, the one that did this best was Google, which used a novel citation mechanism to identify which websites were most popular around specific topics. Google's algorithm was special because it used something akin to "social proof" to determine what was good. It did this by counting links between sites. So if you were looking for recipes for potatoes, it would look for the sites that mentioned *potato recipes*, and rank-order them based on the number of other sites that linked to them. This innovative indexing method, known as PageRank, found a kind of social evidence in the existing connections between sites. It proved far better than any other search engine available at the time. By measuring this social relationship between websites, Google found a quality signal within the obscure haze of the early web. And by serving well-targeted ads alongside each search result, it rapidly became one of the most profitable sites on the internet.

But in the early 2000s, if you weren't searching for something, the internet didn't know how to serve you content. Most content-driven websites still relied upon the traditional model of publishing like newspapers: They paid people to create articles and then served it on a home page. Users themselves needed to poke, prod, and pull at the internet to be served information.

Yet something new was about to turn that relationship on its head.

Connecting Friends of Friends of Friends

When it all began, social media felt like a kind, gregarious, and pleasant place. Friendster, Myspace, and Facebook all emerged between 2002 and 2004, offering something simple: a standardized website just for you. It gave you a profile, a URL, linked you to your friends, and gave you a few basic messaging tools to talk with them. They encouraged people to post highly curated versions of their lives and show off their friendships. There was no news, and there was no feed. They proved to be enormously popular with teens and young adults. Friendster, the first major social networking site, groaned under the enormous load of the traffic it generated.

Soon another competitor came along. Arguably a worse platform, Myspace had less friction in place for adding new friends and expanding connections between users. As Friendster buckled under the traffic of millions of requests, their users began flocking en masse to Myspace, where the servers were kept mostly up and running with little to no performance issues.

However fast the back end was, the front end of Myspace was actually poorly built. As the site took off, one of its main features was actually a security bug that was exploited by users: the ability to personalize your page using code insertion. Some enterprising users learned that they could inject their own CSS code into open text fields, thus changing their backgrounds, adding images, media players, and dramatically modifying the look of their personal profiles. This security bug ended up being the site's most compelling feature—deep customization. Since users loved it, rather than cleaning up the bug, the site's developers made the decision to leave it open. The value of this customization soon became a liability, however, as the wide-open latitude given to each user's individual profile turned the site into a mess of bad design.

Myspace had a wide open social graph and a culture of adding as many friends as possible. Users were encouraged to add strangers to increase their friend count. Every new user was connected to a guy named Tom, one of the site's founders. These two features led to rapid adoption and massive growth, peaking at more than 130 million active users. People competed to add more friends and show their popularity on the platform.[7]

But this frictionless explosion of strangers connecting to strangers didn't last. In 2006, when the company sold to NewsCorp for $580 million, it was already losing its shine. It had begun to resemble a seedy and garish public carnival with no rules, full of unfamiliar faces and ugly design. These were not real relationships, and there was not a binding force in most people's connections. It had grown too big, too fast. Users were ready for something else.

That something else came in the form of a startup called Facebook, launched by Mark Zuckerberg and a handful of his roommates in 2004. The story of Mark and his friends building the world's largest social network has been well told. What's most important is that Facebook won this early battle for dominance of our online lives by making sure these social relationships were real by linking users to college email addresses. Zuckerberg called this web of relationships the social graph, predicting that owning a network of real-life bonds between humans would be valuable.

Within a few years of steady expansion, this prediction would hold true, as hundreds of millions of people found themselves perpetually online in these targetable communities of social media.

The Viral Upgrade

A handful of incremental changes that happened between 2009 and 2012 dramatically upgraded social media—namely the speed of how we as individuals share information. Independently, each of these changes wasn't problematic. Each was a feature set that solved a problem, and each was immediately, measurably helpful to both users and the platform's creators.

When Twitter launched in 2006, its most distinctive feature was its timeline: a constant stream of 140-character updates that users could view on their phones. The timeline was a new way of consuming information—an unending supply of content that, to many, felt like drinking from a fire hose. Later that year, Facebook launched its own version, called the News Feed.

Facebook and Twitter, at slightly different times, added a major innovation: a ranking algorithm that determined which content a user would see first based on predicted engagement—the likelihood of an individual interacting with a given post. This turned the timeline into something more manageable and interesting, a flow of tailored content that you yourself were likely to react to.

The algorithmic ordering of content soon flattened the hierarchy of credibility. Any post by any producer could rise to the top of our feed as long as it generated engagement. Misinformation would later flourish in this environment, as a personal blog post was given the same look and feel as a story from the *New York Times*.

In 2009, Facebook added the Like button, and for the first time created a public metric for the popularity of content on the platform. It was quickly copied by Twitter and became a foundational metric for determining the value of content shared online.

Twitter also made a key change in 2009, adding the Retweet button. Until then, users had to copy and paste older tweets into their status updates, a small obstacle that required a few seconds of thought and attention. The Retweet button enabled the frictionless spread of content. A single click could pass someone else's tweet on to all of your followers. In 2012, Facebook offered its own version of the retweet, the single-click Share button, to its fastest-growing audience: smartphone users.[8]

This single-click Share button turned people into active participants in the distribution and amplification of information. News feeds pushed out bite-size posts to friends and friends of friends. These three primary changes worked in concert together: Curation algorithms used Likes and Favorites to decide what to showcase, basic recommendation algorithms then pushed content to users, who boosted the most engaging content even further by sharing to their networks on their own.

These three changes—social metrics, algorithmic feeds, and the one-click share—fundamentally shifted the types of information that were available to all, and together constituted a sea change in how we parse and process new information. Sitting in our pockets in a shiny new device—the smartphone—together these three inventions fundamentally altered the type of content that we see regularly in our lives. Each played an outsized role in transforming our media ecosystem into this strange, outrageous place that it is today.

These improvements constituted one of the greatest shifts in information sharing since the invention of the printing press, drastically increasing the speed and spread of information we create and consume. Together, these three innovations ushered in a new era we're living in today: the viral era.

The Viral Brain

What happens when we increase the speed of a network? What happens to our shared perception if we're flooded with new information? We begin to reach the limits of our ability to carefully parse knowledge.

A helpful and practical metaphor can be taken from the Nobel Prize–winning work of Daniel Kahneman, whose research with Amos Tversky outlined two key "systems" in our mental operations. System 1, the fast, instinctive, and emotional; and System 2, the slower, more deliberative, and more analytical way of thinking and consuming information. System 1 is predisposed to biases and mental shortcuts that allow us to make snap decisions, while System 2 helps us with complex and nuanced problems.[9]

Fast vs. Slow Thinking

SYSTEM 1	SYSTEM 2
"Fast" Thinking	**"Slow" Thinking**
Reactive	Reflective
Unconscious	Reliable
Emotional	Effortful
Automatic	Deliberative

Both systems are helpful in our daily lives, but System 1 thrives on digital architecture that prioritizes speed and impulsivity. From clickbait to emotionally arresting, outrage-inducing news, the social web was inadvertently built to capitalize on System 1, tilting us toward the reactive, automatic, and unconscious.

The entire architecture of the social web operates with virality in mind, with the specific goal of capturing our attention as fast as possible. With no friction between our neighbors' impulsive thoughts and our own, impulsive fast thinking takes precedence.

The feeling of being inundated by this type of information is one of urgency—everything becomes critical. Our System 1 brains don't know how to parse the emotionally urgent from what is genuinely important.

A casualty of this increase in speed is actually often truth itself. When Seneca the Younger apocryphally wrote: "Time discovers truth," he was identifying a core principle of knowledge-sharing. We still hear his idiom today as "time will tell." In deciphering what is false and what is true, time is a critical partner to accuracy. The longer we have, the more opportunities we're given to filter, assess, and confirm. For this reason, the vast majority of falsehoods we see online are seen quickly, shared quickly, and travel faster than our ability to stop them. We didn't know it, but virality supercharged the emergence of misinformation.[10]

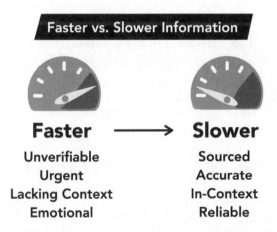

Faster vs. Slower Information

Faster ⟶ **Slower**

Faster	Slower
Unverifiable	Sourced
Urgent	Accurate
Lacking Context	In-Context
Emotional	Reliable

In our media environment, the speed of content has come to matter greatly. According to research led by Soroush Vosoughi and Deb Roy at MIT, fake news spread up to six times faster and more widely than authentic news.[11] It has a viral advantage that's hard for factual information to compete with. When this study was released in 2018, it confirmed that much of the most viral information we see is inaccurate.

In 2022, the study itself was challenged in an amusing illustration of exactly what is so wrong with viral information online. A journalist for *The Atlantic* was confused by another academic's commentary about the original study and stated that its key findings were wrong, summarizing it in a tweet, saying, "I love this so much: Remember the *Science* paper showing that misinformation travels farther and faster on social media than the truth? It was wrong!" This tweet then went wildly viral. But, in fact, the tweet was incorrect. *The Atlantic* journalist then retracted his false claim of the study being false. The article's headline? "Sorry I Lied About Fake News."[12] Vosoughi and Roy's study still holds up, even if a falsehood about it went wildly viral in a poetic example of precisely the issue.

Virality

is the phenomenon of content spreading through peer-to-peer sharing.

Velocity

is the *rate* at which that content spreads.

Velocity and virality are different. Low-velocity content can still go viral: a good book we share with our friends, for instance, or a word-of-mouth recommendation for a film. But high-velocity content is more likely to appeal to our immediate emotions and our System 1 brains.

A Viral Superpower

But way back in the 2010s, when viral content first began flowing into our feeds, it seemed like a gift—a fresh power given to all of us. A magical way for anyone, anywhere to suddenly find fame, acceptance, and recognition, simply based on the merits of their content alone. It appeared then to be the beginning of a virtuous revolution. Free of human gatekeepers, free of stuffy mainstream media powerbrokers, and free of the old guard.

The earliest content that went viral on social media was exceptionally funny. The user-generated flood of videos, articles, and memes that followed were interesting, entertaining, and useful. They began steadily percolating into our feeds, along with a wellspring of stories about people whose lives had been changed by this unexpected fame. Many creators found huge new audiences overnight. Virality felt just, egalitarian, and real—a generational shift in how we present ourselves. It felt as if we could suddenly see the rest of humanity better than ever before.

But implicit in that seeing, we were also now being seen. By using these tools we were entering into a strange contract—a contract of performance and assessment with friends and strangers alike. We were now observed, evaluated, and judged with every post. The invitation and incentive were there for each of us to contribute to the evolving spectacle of social media. And as that spectacle unfolded, we were about to find ourselves shocked by what we would witness.

IN SUM

We began this chapter exploring how we were all born into an explosion of information. We consume vastly more content on a daily basis than at any previous point in our species' history, while our brains' capacity to process it has stayed roughly the same: roughly 120 bits per second.

We tracked through the history of the internet, examining how it began as a social network in its own right: a documentation system of hyperlinks that provided a means of finding collective knowledge. We then learned how a new, highly personal internet emerged—social media—which evolved in fits and starts.

- Social media changed dramatically beginning in 2009, when three key features, *algorithmic feeds*, *social metrics*, and *one-click sharing*, fundamentally upgraded the speed at which we spread knowledge, propelling us into the modern viral era.
- A key principle of social media is that *fast-spreading information tends to be false.* It behaves this way because viral content tends to appeal to what Daniel Kahneman refers to as our System 1 brains, relying upon emotional heuristics and intuitions, rather than System 2, the more deliberative part of our cognition.

We're about to learn just how these tools became so addictive, and the surprising effect they have on one of our most important faculties: our ability to focus.

Chapter 4

The Origin of Our Addictions

**We Would Like to Send
You Notifications**

Notifications may include terrifying
news, subversive alerts, and variable
hits of dopamine.

Don't Allow   OK

I'm going to tell you a few things you already know.

Every time you open your phone or your computer, your brain is walking onto a battleground. The aggressors are the architects of your digital world. Their weaponry are the apps, news feeds, and notifications in your field of view every time you look at a screen.

They are attempting to capture your most scarce resource—your attention—and take it hostage for money. In order to succeed, they need to map the defensive lines of your brain, your willpower, and your desire to concentrate on other tasks, and figure out how to get through them.

You'll lose this battle. You have already. The average person loses it dozens of times per day.

This may sound familiar: In an idle moment, you open your phone to check the time. Nineteen minutes later you regain consciousness in a completely random corner of your digital world: a stranger's photo stream, a surprising news article, someone dancing on TikTok, a funny YouTube clip. You didn't mean to do that. What just happened?

This isn't your fault. It's by design. The digital rabbit hole you just tumbled down is funded by advertising, aimed at you. Almost every "free" app or service you use depends on this surreptitious process of unconsciously turning your eyeballs into dollars, and they've built sophisticated methods of reliably doing so. You don't pay money for using these platforms, but make no mistake—you *are* paying for them, with your time, your attention, and your perspective.

These decisions are not made with malice. They are made behind analytics dashboards, split-testing panels, and walls of code that have turned you into a predictable asset, a user that can be mined for attention.

Tech companies and media organizations alike do this by focusing on one oversimplified metric, one that supports advertising as its primary source of revenue. This metric is called *engagement*, and emphasizing it above all else has subtly and steadily changed the way we look at the news, our politics, and each other.

This addiction to our devices isn't a distinct issue from the problems we're facing in our politics; it's actually the same system.

For the first time, the majority of information we consume as a species is controlled by algorithms built to capture our emotional attention. As a result, we hear more angry voices shouting fearful opinions and we see more threats and frightening news simply because these are the stories most likely to engage us. This engagement is profitable for everyone involved: producers, journalists, creators, politicians, and, of course, the platforms themselves.

The machinery of social media has become a lens through which society views itself—it is fundamentally changing the rules of human discourse. We've all learned to play this game with our own posts and content, earning our own payments in minute rushes of dopamine, and small metrics of acclaim. As a result, our words are suddenly soaked in righteousness, certainty, and extreme judgment.

When we are shown what's wrong in the world, we feel the desire to correct it. We want to share these transgressions with our networks. If we see more problems, these problems *must* have perpetrators who are responsible for them. These enemies are now everywhere, and we feel the need to call them out.

The result is a shift in our collective perception. We see a world under threat: a constant moral assault on our values, a poisonous political landscape, and an abrupt narrowing of our capacity for empathy.[1] These new tools are fracturing our ability to make sense, cohere, and cooperate around the deepest challenges facing our species.

Let's start by unpacking a choice that was made years ago. It was one that most people didn't think about, a simple purchase that most were excited to make. For me, it was a tool that I was personally confident would improve my day-to-day productivity. After deciding to follow an orange-robed monk into the countryside, this was the second-most significant decision of my life.

This seemingly harmless, but profoundly consequential choice was my first purchase of a smartphone. The day I opened this glass rectangle and powered it on, I had no idea that it would change my relationship with my brain.

If you fell asleep in 2007 and woke up in the 2020s, you'd find yourself in an episode of *The Twilight Zone*. Craned necks focusing on glowing rectangles, everyone under some persistent spell. The outside world would be mostly the same, but its inhabitants would be markedly different. Those around you would be speaking an unfamiliar language, obsessed with this luminous box in their hands.

You may remember, hazily, what it was like back in the early aughts. Falling into a book or a long magazine article was easy. Boredom was possible. Long walks by yourself, disconnected from the matrix of urgent knowledge. Not overly worried about missing something important. Not aware of the critical undercurrent of the next looming crisis.

I felt this pain early. My brain chemistry was especially susceptible to this change, particularly vulnerable to the architecture of our phones, our apps, and our feeds. I spent thousands of hours caught within the smartphone-enabled dopamine trap attached to my body. I could feel my daily ability to focus narrowed, excised, dissolved, and diminished as this extraction of my attention became more efficient.

As my attention waned, at first I was confused: My smartphone was supposed to make me more productive, but I was also losing part of my capacity to focus. Why? I studied my own actions like a clinician, trying to fix, manage, and reconfigure my digital environment so it served me, haltingly taking back control.

Years into this journey, I was officially diagnosed with ADHD. It was a painful realization, and one that began to clarify why my struggles were so intense. I was attempting to cope with the increasing informational demands on my brain, and my brain was particularly vulnerable. ADHD has become one of the more commonplace cognitive disorders of the smartphone era, suggesting that there is a link between our use of these tools and our collective inability to focus.[2]

As the writer and productivity expert Ryder Carroll has described, having ADHD in the smartphone era world is much like trying to catch the rain with your hands.

> You step outside and you bring your attention to the darkening sky. The first drops fall. You catch one, then another. Soon the storm picks up, and the rain falls faster. You miss a drop, then another. Soon there are so many things raining down on your attention that you don't know what to focus on. Do you go for the ones coming from a distance, or the ones close to you? The more you frantically deliberate, the more you miss.[3]

For me, it was a slow, painful process of reconstructing my attention. Over a decade I personally built an elaborate Rube Goldberg–type machine to keep myself on track. My personal-focus machine involves a half-dozen browser extensions, news feed blockers, meditation rituals, VPNs, and productivity timers. Each helps me capture a small additional fraction of my attention that would otherwise slip into an infinite digital rabbit hole.

These strategies aren't always successful. I still can't escape the use of these tools in my professional and personal lives, and you're most likely in the same position. My connection to my income depends upon the usage of these instruments. The success of this book depends upon them. No notable figure can live entirely separated from social media without concessions. I must play the game to survive and achieve, and you probably do too.

If you do meet someone without a smartphone (and, I assure you, they still exist), it's a bit like meeting someone with an obscure medical condition. You're curious about their life: You want to know, are they okay? You're proud of them for overcoming their hardships, but you wouldn't be able to do what they do.

Yet we may be the ones who are strange. I can attest that a chunk of my personal agency has been lost since the advent of smartphones. I can measure the loss in weeks, months, and years of my life. I know I'm not alone.

A 2018 study found that 63 percent of smartphone users say they've attempted to limit their usage, with only half of them feeling like they were successful. In 2022, Americans spent nearly five hours a day on mobile devices. For many of us, that's nearly a third of our waking life. We have let them burn into our sleep, we have allowed them to eat our relationships, and have made ourselves sick with frustration, anger, exhaustion, and FOMO.[4]

We know it, but we cannot stop.

Attention is zero-sum. We unfortunately have only so many hours in our waking lives, and when we use our attention, it's gone forever. It's a nonrenewable moment of our finite existence lost. And increasingly it has become our most scarce resource.

What does the world look like when our attention is pilfered, overwhelmed, and extracted from us? What happens when we collectively lose our ability to focus? What are the net costs to the economy and society? These questions were seldom asked by the early architects of our digital environment.

Trying to weigh the trade-off between what these tools are doing *for* us versus *to* us is hard to parse. In spite of the wonders of our seemingly magical connection, parts of our lives are lost. And for many of us, it wasn't really a choice.

Once enough humans begin using a thing, it becomes the default and expected norm. Today, you can't easily function in society without a mobile device. Depending on your industry, that may or may not include the usage of social media. Google, Facebook, and Instagram are critical tools for marketers and any business working on the internet. Use of Twitter is a non-negotiable job requirement for most journalists. Politicians must use social media to reach their constituents. These tools have come to touch almost every part of our public lives. We've adapted to these tools so rapidly that the question deserves to be asked: Did we choose this weird world? And if we didn't, who did?

The Two Sides of Choice

On a brisk New York evening in 2014, I climbed into an Uber home from a party in Midtown with a handful of friends, on my way back to my apartment on the Lower East Side. I found myself sitting next to a bright-eyed young man with reddish-brown hair and a quiet manner. As we chatted, we discovered we had grown up just miles from each other in California, and had both lived and worked in tech in San Francisco at the same time. I mentioned I was in the early stages of researching this book. He told me he was in the beginning stages of diagnosing a problem he saw unfolding at Google where he worked. We decided to meet up the very next day for lunch.

His name is Tristan Harris. Over Vietnamese food the following afternoon, he told me how, less than a year earlier, he had realized that there was a strange mind-set among product designers in Silicon Valley. He saw a widening gulf between the incentives driving the designers and engineers who created the products, and the best interests of their millions of customers. Specifically, he noticed that the drive to maximize internal metrics like "time on site" was increasingly in conflict with what was best for users' attention. Design tweaks like infinite scrolling, intrusive alerts, and other so-called attention hacks were being widely used in the industry to keep people hooked on their products.

He had created a presentation entitled "A Call to Minimize Distraction & Respect Users' Attention" that he shared internally at Google. The 141-slide deck made a case that the company had a fundamental responsibility to ensure its users didn't spend their days buried in devices at the expense of their quality of life. After he shared it with a few people, it went viral within the company, and was viewed by thousands of Google employees.

Tristan was articulating a big new idea—one that had yet to be fully fleshed out. He was concerned that distraction was at the core of a fundamental new dilemma for the tech industry. Not a business problem, but a philosophical one. He could see that "choice" itself was becoming a squishy concept when it came to the usage of these tools.

Most entrepreneurs and product designers adhered to the basic tenets of liberal economics: the idea that individuals choose the best products in the marketplace, and subsequently reward entrepreneurs that create them. Build a better mousetrap, it's said, and consumers will choose it over your competitors.

But in practice, many Silicon Valley product designers were also attentive students of *behavioral* economics, a field that recognizes that consumers are often predictably irrational, and that there are very clear psychological triggers to make people act in a certain way, regardless of their preferences.

Behavioral economics shows how to bend human decision-making. A classic example of this is the irrational market incentives of a slot machine—a game that uses irregular payouts to hack the pleasure center of players—with the rush of occasional wins netting out to a loss over time. Slot machines use a strategy known to behavioral scientists as *intermittent variable rewards* to get players hooked, something that Tristan saw product designers emulating in the design of news feeds and push notifications.[5]

Tristan believed that this was about far more than distraction. He sensed that many of the features of these tools were beginning to pull us away from ourselves and into an impulsive advertising-driven dystopia. In his opinion, designers, engineers, and advertising companies alike were preying upon our attention at a steep cost. These costs include degrees of human agency—our literal free will.

Tristan became a close friend, and over the following years his message exploded. After leaving Google in December 2015, he launched a movement known as Time Well Spent, which later became the seed-crystal of a movement to reform technology to serve human interests. A few years later, he launched the Center for Humane Technology with the goal of mobilizing support for tech that aligns with human values while protecting people's agency. By 2016 he had become a mainstage TED speaker and was in high demand in the thought-leadership circuit. At that moment, Tristan had carved out a very distinct side of the argument against exploiting our attention.

A few years later, I found myself facing the other side of this issue, quite literally, at a dinner party hosted by John Stossel, a prominent New York journalist and libertarian. John, in the interest of provoking thoughtful conversation, regularly hosted dinner salons focused on controversial issues, believing that through the process of putting conflicting ideas together over dinner, the best perspectives might be revealed through kindly debate.

The man I faced on the other side of this issue was Nir Eyal. Nir has written several books, including *Indistractable: How to Control Your Attention and Choose Your Life*, which focuses on how to make people immune to the tools and tricks of attention capture. If this sounds positive, it might be viewed as a sort of penance for Nir's first book, the bestseller *Hooked: How to Build Habit-Forming Products*, which he used to launch a successful consulting business for companies looking to maximize the capturing of human attention. Nir was a pioneer in disseminating the strategies of gamification and addiction, and his work was widely adopted in Silicon Valley.

Nir has become something like an Anti-Tristan, telling people that these assumptions and fearmongering about human attention and agency are, in his words, bullshit.[6] That the process of deferring responsibility for our choices to tech companies is, in his view, a terrible lie. He believes that removing ourselves from the decision-making process is a ridiculous short-changing of our personal agency. In his view, it's disempowering and unfair to the individual: a story about how we spend our time that takes away our freedom.

Nir often quotes the French philosopher Paul Virilio: "When you invent the ship, you invent the shipwreck," referencing the calamitous by-products of every new technology. As a foil, Nir is very effective, and he makes strong points. I can see the logic in both arguments. But he's wrong in believing that we're not facing a fundamental crisis at the scale that Tristan has described.

Over the years, Tristan and Nir, through very different means, have come to hold two separate corners of the same truth. As with many debates, a larger story emerges from comparing their central arguments. We *are* undergoing a broad systemic shift where our attention is being targeted, extracted, mined, and plundered better than it ever has before. But we're likely to manage this crisis better if we recognize we do have some control and take the steps to change it. If we empower people with a better understanding of what these tools are doing, we can modify our behavior, and we can demand more from the platforms themselves.

IN SUM

As we learned in this chapter, free apps that monetize through advertising often use hidden tools and tricks for capturing and maintaining our attention.

In order to create addictive products, designers have introduced behavioral strategies such as *intermittent variable rewards* to keep us using them. In addition, they use algorithms that maximize for a metric called *engagement*.

When we don't pay for apps with money, we are paying with our attention. Attention is a finite cognitive resource.

Most importantly, we learned that these tools play a role in manipulating something much deeper: our sense of personal choice. For the designers of our social tools, human agency itself is a flexible and malleable attribute.

Advocates like Tristan Harris argue that we are in a crisis, and the onus is on the industry to design more humane technology. Writer Nir Eyal believes individuals carry the full responsibility of overcoming their distractions. This debate shows the difficult nature of assigning blame when the core asset being extracted is attention. Attention is central to our sense of personal agency, and when it's extracted, we often blame ourselves. But we can take more responsibility.

Next, we'll dive deeper into how social media has inadvertently come to shift not just our own attention as individuals, but the attention of millions of people all at once.

Chapter 5

Pushing the Trigger

In February 2016 I was hired to consult with a boutique design and strategy firm in New York City. Our client was the executive team of one of the world's largest news conglomerates, which I will call NewspaperCo, a significant owner of local newspapers. The massive publisher had seen continual declines in advertising revenue and was struggling to figure out how to make its business work in the new social-media-attention marketplace, where a majority of new ad spending had migrated. We were hired to work with the company's board of directors, and it was my job to identify the specific problems with their news business.

Upon digging in, what I found was a legacy institution in fiscal free-fall. NewspaperCo had watched its advertising revenues decline by double digits over a decade as its customer base was aging, and the dozens of local papers it owned were being crushed by dwindling subscriptions and huge debt obligations.[1]

The elephant in the room was something they weren't sure how to handle: social media's influence on their readers' attention. I had expertise in this, and I felt like a double agent. While my day job consisted of helping to analyze why NewspaperCo was failing, I had another client: a startup that had hired me to take advantage of a bit of that failure—a social media tool that was hacking people's attention.

This startup's mission was commendable. It was dedicated to getting people to care about social causes that they otherwise would have ignored. It was an advocacy tool, helping to increase the prominence of various nonprofit issues on social media.

Our goal was to amplify messaging for particular campaigns, creating a cascade of viral support for a cause. If traditional newspapers were meant to objectively reflect the stories of the day, we were working to make stories about social issues objectively sensational.

Our collaborators on these projects included the stars of popular film franchises like *The Hunger Games*, as well as other major celebrities. We were testing ways to set off a chain reaction of viral posts about a topic, with the purpose of funding public health projects like Ebola relief, treatment for preventable diseases that cause blindness, ending homelessness, and increasing access to surgical procedures in developing countries.

Our tools were powerful: We used Facebook's ads manager coupled with several custom integrated ad products. These allowed us to run targeted advertising campaigns that found signals inside Facebook's audience. We could hypertarget a market with a message and get an instant cost per click—guaranteeing that some version of our message would find eyeballs.

In a way, we were virologists working inside a lab that Facebook had built. Our objective was to engineer content that would spread the farthest in the service of the causes we cared about most. This process is known as optimization—a tailoring technique, taking a piece of content and turning it into the most viral version of that idea. The goal was to make things expand beyond our network into third- and fourth-degree networks.

Harnessing virality was still an emerging science. If you were sharing content, most people in your personal audience would share a thing because they liked you. To make *their* audience click and share a thing required a very specific set of steps. To get third-degree sharers to click *and* reshare a post, that was a real challenge.

This was a technique that was pioneered several years earlier by the news site *Upworthy*, one that briefly made it the most popular site on the internet in 2012. Around that time, *Upworthy* created a slide deck titled "How to Make That One Thing Go Viral," which soon went viral itself among Facebook optimizers.[2] It became a bible of sorts for social media marketers and content creators across the internet trying to maximize the attention they could capture online.

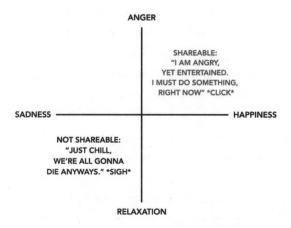

Adapted from *Upworthy*'s slide excerpt titled "Why the Hell Do People Share?" Source: *Upworthy*

They pioneered a new type of split testing, one of the strongest tools in the optimizers' tool kit. Their method involves writing at least twenty-five different headlines for the content you're trying to promote. With each headline, you try to make people upset while keeping them entertained. Change the title, the subtitle, the photo, or the inline text, and show them to a split audience. Each variable would become a test of maximum engagement. Over time, these tools pushed more traffic to the ad that garnered the most clicks—the winner of the test. We could then use that to change the copy of all posted articles.

At this moment, Facebook was king of traffic on the internet. The company had just surpassed one billion users and drove a majority of all news traffic world-wide.[3] Most other social media platforms had much smaller audiences and much smaller viral coefficients (the number of new consumers who came from an existing consumer of content). Twitter was tiny in comparison. Reddit was hard. Tumblr required finesse. But Facebook was easy. It owned the exponential referral cycle—the users that would share onward to their friends, who would share with their friends, and so on. At that moment Facebook was a viral colossus.

Facebook's tools helped us identify language that might hijack our audience's attention, pulling them out of a mindless scroll and into a sudden click. But Facebook's algorithms seemed to have a strange agenda. And in our process of optimization, we were seeing strange, dark patterns beginning to emerge.

What we found was that using emotional language in our posts led to *far* more engagement and far more shares than other types of content. There was a natural reservoir of interest in items that evoked both emotional and moral reactions. Internally, we referred to this as *affective engagement.* Unfortunately, it had a clear bias toward negative emotions.

More recently, researchers at New York University, the University of Giessen, and ETH Zurich studied *Upworthy's* data set and were able to confirm what we were seeing in our own tests:

> Our analyses revealed that negative words had a positive effect on click-through rate, while positive words had a negative effect on click-through rate. This suggests that a larger proportion of negative words increases the tendency of online users to access a news story (and vice versa for positive words).[4]

What was more worrying was that we could see this trend everywhere on Facebook. The strategy seemed to be directly transferable to news stories, political campaigns, propaganda, and any type of content people wished to promote. If you wanted to capture attention, the data was clear: get people upset. Negativity paid dividends.

Meanwhile, back at NewspaperCo, the company was still struggling to figure out how to do anything meaningful with their content on social media. Their stories were still good, mainstream, credible, with high standards of journalism. But they could not compete with what was happening on Facebook. Much of their audience had gone online and left them behind.

Over the course of several months, I watched with curiosity and horror as many of the strategies my friends and I were employing to hack traffic became a new standard for news delivery and dissemination. Many publishers would find something problematic online, repackage it with an outraged headline, and turn it into a viral article for clicks.

These tools became available to anyone with a budget and a little know-how, at the expense of factual accuracy and journalistic integrity. NewspaperCo couldn't keep up, and they were getting increasingly desperate. Old-guard news organizations across the board didn't know what to do.

A Race to the Bottom of the Brainstem

Today, you're unlikely to see any headline that hasn't gone through some substantial split test to optimize engagement. With nearly every piece of journalism you see on the internet today, the person who writes the headline is rarely the author of the story itself. As *Fusion* editor Felix Salmon wrote that year in Nieman Lab, "The amount of time and effort put into packaging a story can significantly exceed the amount of time and effort that went into writing it in the first place."[5]

Sensationalized headlines are stickier and gain more traction. They propagate faster and drive more traffic than their less hyperbolic counterparts. A certain breed of news producer was the first group to adapt to this powerful new trend, and they took the platform by storm. A sample of top performing word series from a 2017 study of 100 million headlines from this early era showed just how dramatically this landscape had changed.

1. Tears of joy
2. Make you cry
3. Give you goose bumps
4. Is too cute
5. Shocked to see[6]

Headline packaging quickly became a way in which every news story was contextualized, or repackaged, specifically to garner the maximum quantity of attention.

A/B TESTING IN SUMMARY

THIS IS BOB

NEWS ITEM!

BOB DID A THING.

This news story will be changed to receive more clicks by tweaking the title, subtitle, and photo to test the reactions of the audience.

@TOBIASROSE

When you measure the ones that get the most clicks, writing a headline can then be abstracted into a game with the goal of capturing as much attention as possible.

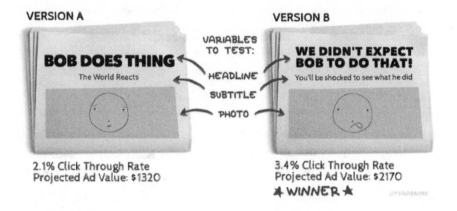

With these tools and a small amount of creativity, a factual story can become provocative or sensational simply depending on how the headline is written.

Unfortunately, most people who see these posts on social media don't click through to read the articles themselves. It's not uncommon for users to interpret the headline itself as the story, even if it doesn't resemble the original event.[7]

It's not hard to see how these strategies might be used to turn the news hyperpartisan, divisive, and outrageous. As the former head of content at a major millennial-focused publisher told me that year, "It's not our job to challenge political opinions. It's our job to ride your politics as far as we can."[8]

Savvy publishers know that partisanship is a powerful driver of engagement. Generally, people prefer to click, comment, and share things that make them feel good, and stories that affirm their beliefs—or outrageously challenge them—are particularly engaging. What's more, stories that trigger our moral emotions will make us click. To many news organizations, violations of our values are excellent opportunities for engagement and monetization.

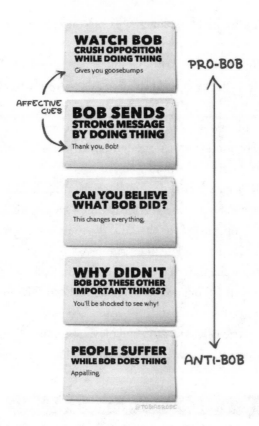

Virality Becomes News

Sitting between this startup and NewspaperCo, I could tell that something fundamental was shifting in how news companies of all types—not just online news—were changing their strategies. Traditional journalism was changing rapidly and doing anything it could to stay profitable.

This transition was rapid. By the first half of 2016 a new trend toward extreme optimization had become widespread in cable news as well. This could be seen directly in the data. During that year's election cycle, a traditionally centrist news organization, CNN, began to shift its coverage. By the end of the election season, it had made more than a billion dollars in gross profit above the previous year driven primarily by advertising attached to news about one particularly outrageous candidate: Donald Trump.[9]

And Trump, too, had exploited this new attentional marketplace. The year 2016 was far from the first time he explored running for president. In 1987, 2000, 2004, and 2011, Trump publicly considered a bid for the nation's highest office. In 1999, he entered the race as a Reform Party candidate, testing his platform and evaluating the response, ultimately deciding he couldn't get the traction necessary to win. After his unsuccessful run in 1999, *Newsweek* noted, there simply wasn't enough anger in the country to propel an independent candidate like him to victory.[10]

His ambitions hadn't changed much in the three decades prior to 2016. What was different about those previous years? One key distinction was this: The media wasn't optimized for the kind of emotional urgency necessary to provide coverage for a candidate like Trump.

This was the mechanism that came to define the 2016 campaign: The more outrageous his words, the more coverage he received. The more coverage he received, the more viable his candidacy became. The analytics firm mediaQuant estimated that between October 2015 and November 2016, Trump received $5.6 billion in "free" earned media from this strategy, three times his nearest rival.[11]

Snapshot of primary season media coverage, March 14–23, 2016, Source: Ev Boyle/USC Annenberg

Historically, for better and for worse, mainstream media had been a sort of kingmaker for aspiring politicians. It had the luxury of choosing where to focus the narrow spotlight of our collective attention. But when their audience went to social media, they had to follow them there. And on social media, outrage brought the eyeballs. Having media platforms cover you is a massive advantage in politics. In any election, one of the principal challenges is rising above your competitors and getting noticed. Doing something outrageous gets you noticed. These stories about candidates traveled faster and further on social media than anywhere else. Facebook and Twitter, like CNN, saw massive traffic and revenue spikes thanks to the sensationalized news propagated on their platforms and the attention they captured.[12] Trump's ideology, attitude, and statements played upon anxiety about global threats. The legitimacy of his candidacy was partially dependent upon many of those threats being perceived as real.

Broken News

If you're still questioning how or why a mainstream news company might have sacrificed its integrity for the purpose of chasing fleeting clicks, let's talk about how news is curated.

Most news organizations used to have an old-school editor who would make calculated decisions about what their audience wanted, what needed to be reported, and what was newsworthy. Today, that editor is augmented by algorithms. This editor uses a suite of tools to help them make decisions and target you more effectively. Twitter now plays a critical role in how journalists source news.

The discovery of these tricks for extracting attention in the news can be likened to the discovery of hydraulic fracturing as a way to extract oil. It created a massive boom of emotionally arresting media, flooding our airwaves and news feeds. Also, just like fracking, without any environmental safeguards, the by-product is toxic and flows into our collective conversation, where it pollutes common discourse.

In the United States, nearly every journalist uses social media. Editors use it every day to decide how to allocate their coverage of critical issues. The

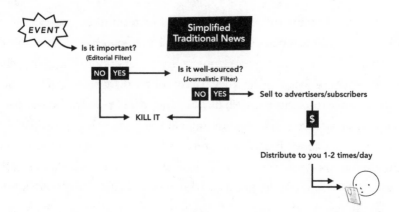

industry of journalism has been consumed by the social media news feed. An economic dependency has taken hold, as those who are responsible for sourcing truth have become professionally and personally addicted to these tools.[13] Many journalists even see tweets as equally newsworthy as headlines from the Associated Press.[14] This vastly increases the risk of bad ideas, fringe content, and false news becoming amplified.

This is a painful open secret in the news business because today the hidden governors of our information system have become algorithms. They're built by humans to capture attention at almost any cost. And increasingly, that cost is our civility, decency, and measured discourse.

NewspaperCo was struggling specifically *because* they weren't participating in this new race to the bottom. By the time I left the project, it was clear that they were desperate to stay alive, and everything was on the table.

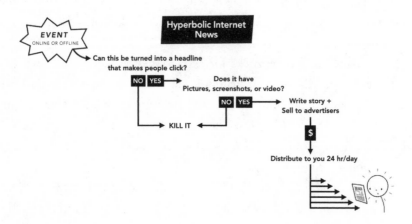

Some time after I departed my consulting work with NewspaperCo, this new brutal pressure would be realized with examples across the industry. In 2021, one of the most recognizable and moderate newspapers in the US sold their front page to an ad campaign that broke one of the core tenets of journalism: *Don't say things that aren't true.* *USA Today* wrapped its cover in a full front-page story declaring "Hybrid babies born across the US" coupled with two fictional articles, in an unabashed promotion for a Netflix show called *Sweet Tooth.*[15]

The bizarre competitive race to capture attention in the digital sphere had come to fruition, forcing mainstream newspapers to adopt drastic measures, at the expense of journalism and trust. For those companies unwilling to play the viral game, few options were left. In a few short years, as our eyeballs had shifted to social media, the race to capture our attention had broken the news.

Moral Panics and Real Panics

An illustration of just how strange our media environment has become can be seen in the way one particular news story exploded across the American media landscape. Since COVID-19, we've forgotten the other pandemic that swept through our lives. The first modern disease outbreak was of a different kind—a pandemic of panic. It was a public health emergency divorced from common reality. In many ways, it was the dress rehearsal for COVID-19.

One evening in late October 2014, a doctor checked his own pulse and stepped onto a subway car in New York City. He'd just returned home from a brief stint volunteering at a medical mission overseas and was heading to Brooklyn to meet friends at a bowling alley.

He was relishing this break. Earlier that day he'd taken a run around the city, grabbed coffee on the High Line, and eaten at a local meatball shop in Chelsea. When he woke up the next day exhausted with a slight fever, he called his employer to tell them he was feeling sick.

Within twenty-four hours, he would become the most feared man in New York. His exact path through the city would be scrutinized by hundreds of people. The establishments he visited would be shuttered. His friends and fiancée would be immediately put into quarantine. Dr. Craig Spencer had contracted Ebola while he was treating patients in Guinea with Doctors Without Borders.[16]

Fortunately, he was, in reality, not contagious until long after he was put into isolation. He followed protocol to the letter in reporting his symptoms and posed no threat to anyone around him while he was in public. He was a model patient, a fact readily shared by medical experts, the CDC, and hospital officials.

This didn't stop a media explosion declaring an imminent viral apocalypse. A frenzy of clickbait and terrifying narratives emerged as every major news entity raced to capitalize on the collective Ebola panic.

Though the physical damage done by the disease itself was small, the hysteria, traveling instantly across the internet, shuttered schools, grounded flights, and terrified the nation.

Social media exploded around the topic, reaching six thousand tweets per second, leaving the CDC and public health officials scrambling to curtail misinformation spreading in all directions. The fear traveled as widely as the stories reporting it. The emotional response and the media attached to it generated billions of impressions for the companies reporting on it.[17]

Those billions were parlayed directly into advertising revenue. Before the hysteria had ended, millions of dollars' worth of advertising real estate attached to Ebola-related media had been bought and sold algorithmically to companies. The terror was far more contagious than the virus itself, and it had the perfect network through which to propagate—a digital ecosystem built to spread emotional fear far and wide.

The Ebola scare was the result of *engagement optimization* gone awry. This type of sensationalism in the news wasn't novel. But the trend of increasing media coverage of threats has steadily distorted our perception of reality at a high level.

More recently, concerns about the US crime rate have followed a similar pattern. Though there has been a sharp uptick in murders in recent years following COVID-19 and the 2020 protests against the murder of George Floyd, violent crime is still far lower than it was in the decade previous. A 2022 report from Pew Research concluded:

> In 2021, the most recent year with available data, there were 16.5 violent crimes for every 1,000 Americans ages 12 and older. That was statistically unchanged from the year before, below pre-pandemic levels and far below the rates recorded in the 1990s, according to the National Crime Victimization Survey.

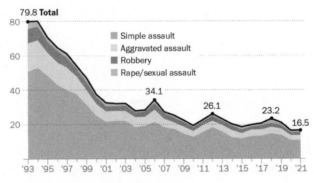

Federal surveys show no increase in U.S. violent crime rate since start of the pandemic

Violent victimizations per 1,000 Americans ages 12 and older

Note: Data for 2006 is not comparable to other years.
Source: U.S. Bureau of Justice Statistics.

PEW RESEARCH CENTER

Source: Pew/Gallup

The murder rate during this period has increased dramatically, by some accounts at the fastest rate since statistics were available. As Pew stated in the same report, "Both the FBI and the Centers for Disease Control and Prevention (CDC) reported a roughly 30% increase in the US murder rate between 2019 and 2020, marking one of the largest year-over-year increases ever recorded."[18]

This trend is very concerning, and deserves our attention, particularly if it continues upward. But even with this dramatic increase in the national homicide rate, murders remain well below levels seen in the 1980s and 1990s. And murder remains the least common type of violent crime overall. (And an even smaller fraction of mortality rates, with the US murder rate in 2020 71 percent below the mortality rate for drug overdoses, which went up over the same period.[19])

U.S. violent and property crime rates have plunged since 1990s, regardless of data source

Trends in U.S. violent and property crime, 1993-2019

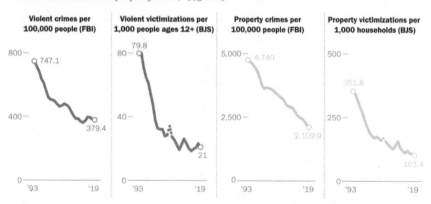

Note: FBI figures include reported crimes only. Bureau of Justice Statistics (BJS) figures include unreported and reported crimes. 2006 BJS estimates are not comparable to those in other years due to methodological changes.
Source: U.S. Bureau of Justice Statistics (BJS), Federal Bureau of Investigation (FBI).
PEW RESEARCH CENTER

Our media system informs our perspective of threats. A focus on crime in news reporting doesn't just change our opinions on crime in general—it makes us feel far more threatened than we should be. For most of us, perceptions are reality. When we see the world as a dangerous place, it changes our behaviors and our attitudes, regardless of the actual threat.

Americans tend to believe crime is up nationally, less so locally

% of U.S. adults who say there is more crime in ____ than there was a year ago

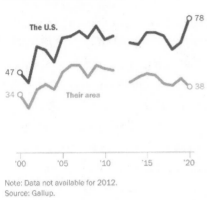

Note: Data not available for 2012.
Source: Gallup.

PEW RESEARCH CENTER

For the better part of the last two decades, an especially critical example of this was terrorism, which, for many of us who lived through 9/11, felt like a real threat to our safety. In the mid-2010s, it was easy to assume that terrorism was one of the top causes of death worldwide.

Yet during this period terrorism-related homicides were a tiny fraction of the overall homicide rate, particularly in the United States. There was a profound asymmetry in coverage of terrorist attacks versus other types of homicides, as illustrated by a two-year sampling of front-page stories collected from the *New York Times* during that time.

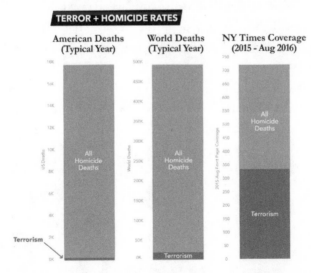

Source: Nemil Dalal, 2017, Pricenomics. Chart shows sampling of data and stories defined as "Islamic terror" compared to overall homicide deaths. Note that terrorist attacks by non-Muslims are regularly not referenced as terrorism by news outlets in initial reporting, though this is steadily changing.

Terrorism is a powerful emotional event, one that seems to insult the very foundations of civil society and human dignity. There are many legitimate reasons for us to be disgusted by such attacks and for us to cover and discuss them publicly.

Yet this is the uncomfortable truth of terrorism's prominence in our lives: We've built an instant distribution system for its actual intent—terror. The fear of terrorism far outstrips the likelihood of it happening to us or anyone we know. More ominously, the excessive coverage of these attacks is often the exact desired outcome of those who commit them.

For example, in 2014 the so-called Islamic State (ISIS) took advantage of this hyperbolic media ecosystem during its rapid rise to prominence. Understanding it was fighting a battle for attention, ISIS prioritized its brand just as much as its military efforts, building a well-funded media wing to push boundaries and exaggerate its exploits as winning, sustaining, and growing. This effort to dominate the media narrative through horrifying acts built ISIS into a principal threat to the West, despite the fact it had a fairly small standing army, limited resources, and essentially no international support.[20]

This exploitation of global media coverage allowed ISIS to use its narrative to recruit fighters from around the world to both Syria and Iraq, as well as inspiring attacks by disaffected individuals with no formal ties to the actual organization.[21]

ISIS and entities like it knew they were battling for attention, and learned how to use the unorthodox weapon of our own media to gain prominence.

The unfortunate truth is that a terrorist attack, a horrifying massacre, or even just a visceral threat—each of these will make real money for a company that sells news.

The media has become a spotlight shining on these individual stories, casting a huge shadow that's far more terrifying than the actual events.

Anecdotes Help Us Remember but Not Understand

Our ancestors evolved in environments that gave us our emotions. As the biologist Daniel T. Blumstein has written, "Fear, honed by millions of years of natural selection, kept our ancestors alive." These emotions helped us *feel* threats to our safety. They helped us *feel* what was wrong in the world around us. Yet our ancestors had a fraction of the inputs we have today. And today we're still stuck with their emotional machinery. This machinery doesn't handle complexity well.[22]

For our paleolithic ancestors, a story about a person may have been all we needed to make a decision about the future. Clustered in small tribes, we didn't have use for models of thought that accounted for occurrences outside the handful of personal stories that could be shared. We would rarely ever interact with people we didn't personally know. A tale about an individual's experience was enough to know what you needed to know to survive.

Hearing a number of stories about nearby wild dog attacks, we might learn to be afraid of wild dogs. This can be seen today in a phenomenon that psychologists call *availability bias*. It is a shortcut for our brains, which makes us assume, "If it comes to mind easily, it must be true."[23]

Since available information was our primary indicator of probability, our brains likely evolved this mental shortcut to help us know what to expect from the world. This shortcut was overly pronounced with threats because the advantages of being afraid of things that might kill us greatly outweighed the costs. For our ancestors, dying was much worse than just being overly cautious. But we no longer live in these isolated, tiny pockets of connection. We now share a vast and unwieldy network of knowledge, news, and responsibilities with millions of other humans. Because of this broad network of exposure to news, we struggle to understand the proportionality of negative events.

This has real-world consequences when we think about prioritizing resources to tackle particular issues. If we want to save lives, we intuitively struggle to understand the difference between single events and broader trends, small problems and large problems. In most people's minds, explaining that sixty thousand people die from malaria every year is a sad and abstract statistic. What's worse is that in our minds that is almost indistinguishable from six hundred thousand people dying (the real number).[24]

But it has far more tangible outcomes when we think about the kinds of feelings that inform our personal and political worldviews. Journalists (and politicians, for that matter) have learned, through decades of trial and error, that two types of sentiment hack our brains better than almost any others: the things that make us afraid and the things that make us angry.

The old media adage "If it bleeds, it leads" is an example of this type of news disposition. And in the modern digital media environment, "Enrage to engage" is becoming just as relevant.

IN SUM

In this chapter we explored something spontaneous and profound that happened this past decade when the news met the internet: Social media became the dominant driver of traffic to news sites globally, while at roughly the same time the bottom dropped out of the newspaper business. Facing double-digit declines in readership, viewership, and print revenue, media companies began using a new strategy to guarantee their content would capture our gaze, using headline packaging and selecting for stories that already garnered traffic online.[25] As a result of these changes, the news business now operates with a clear principle: *Virality is easy news.*

Social media tends to prioritize negatively valenced content. Many of the narratives that capture the most attention are ones that make us afraid and fearful. This disproportionate news coverage of threats has come to distort our perceptions of things like crime rates. While this has been true since long before the advent of the internet, the unique design of social media has tightened the connection between outrage and profit more than ever before. As media companies feel more and more pressure to compete with social media to stay in business, it has ushered in an era of content that has become what Tristan Harris has referred to as "a race to the bottom of the brainstem."

In part I, we have tracked through the early days of the internet, from early euphoria and optimism, onward to a slow and steady recognition that *something* was going wrong. We've explored two key forces behind this shift: increases in velocity and virality, and the accompanying trend of news organizations being forced to optimize for it.

Now, as we enter Part II, we'll begin to unpack the core elements of exactly how social media operates *through* us. These elements include algorithms, social metrics, and moral emotions. Together, these forces generate strange new outputs for society: context collapse, cancellations, and widespread chain-reactions of public outrage that are impossible to ignore.

We'll begin by learning the unexpected way one very strange and peculiar organism has come to reproduce and thrive in the margins of our outraged digital environment: memes.

PART II

Powering the Machine

Chapter 6

Black and Blue, White and Gold

In early February 2015, a week before her daughter's wedding, Cecilia Bleasdale snapped a picture of three different dresses she was considering buying while out shopping. She texted the photos to her daughter to ask her opinion. "Which one is your favorite?" the daughter replied. The mother said she preferred the third one: "The blue-and-black one." Confused, the daughter told her mom that she might need to get her eyes checked. *Mum, that's clearly white and gold*, she thought.

A few days later at the wedding, after seeing the photo of the dress, the musician playing at the event was flummoxed. The wedding singer Caitlin McNeill saw the dress as white and gold—but when she looked up at the mother of the bride, she could clearly see the dress was blue and black. Asking her bandmates to weigh in, half of them saw the photo as obviously blue and black, and half saw it as obviously white and gold. Hours later, the band was so engrossed in their argument and enraptured by the photo that they almost missed their performance.[1]

Soon after the wedding, McNeill posted the photo to her Tumblr page with the caption: "guys please help me—is this dress white and gold, or blue and black? Me and my friends can't agree and we are freaking the fuck out."

The whole internet was about to freak out, too.

Back in New York, Cates Holderness, an employee at *BuzzFeed* who managed the site's Tumblr account, saw the message from McNeill asking for help in resolving the dispute. She showed the photo to other members of the site's social media team, who immediately disagreed about the dress's colors as well. Confused, she created a simple *BuzzFeed* poll, posted it to Twitter, and left work to catch the train to her apartment in Brooklyn.

By the time she got off the subway and checked her phone, she realized something very strange was happening. "I couldn't open Twitter because it kept crashing. I thought somebody had died," she said. "I didn't know what was going on."

Holderness had unknowingly stumbled upon one of the most viral photos in history. By that evening, the page had set a new record at *BuzzFeed* with 673,000 concurrent visitors. At its peak, Tumblr's data director said the page received 14,000 views a second—840,000 views per minute, vastly outstripping all other Tumblr content. The hashtags related to the dress would peak on Twitter at 11,000 tweets per minute. Propelled by social media, the controversy had captured the internet's complete attention.[2]

What was it about this particular image that generated so much intense fascination? The way the dress went viral globally shows social media's core engine at work. The dress captured *strong* emotional certainty from its viewers and provoked extreme emotional reactions from an inherently ambiguous source, particularly the incredulity that others *could not* see the dress in the same way as their peers. The controversy provided the engagement. Then the controversy *about* the controversy continued its spread.

Three months after the image went viral, *Current Biology* published the first large-scale study on the dress, and found that among the nearly two thousand participants in the study, 30 percent perceived it as white and gold, while 57 percent saw the dress as blue and black. Another 11 percent saw it as blue and brown, and 2 percent reported it as "other." When the dress was shown in artificial yellow lighting, the researchers found that almost all respondents perceived it as black and blue, while if the simulated lighting was blue, the respondents saw it as white and gold. Another study found that people who tended to wake up early were more likely to perceive the dress as white and gold—as they were used to seeing items lit by morning light. Night owls, on the other hand, were more likely to see the dress as blue and black because they were more likely to interpret it as being under synthetic lighting.[3]

The photo is an example of an *ambiguous image*. Most of us are familiar with them from childhood in the form of novel optical illusions that can be interpreted in two or more ways. Usually, ambiguous images are visibly

| -30% BRIGHTNESS +40% CONTRAST | ORIGINAL | +40% BRIGHTNESS +40% CONTRAST |

With adjusted black-and-white images of the photo, it's easier for our brains to switch between light-colored and dark-colored versions of the dress.

interchangeable in that we can switch between them easily as we stare at them. The image below, for instance, can be a rabbit or a duck, and our brains can easily move between their relative forms.

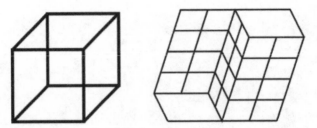

Rabbit/duck and geometric optical illusions. These are two examples of ambiguous images where viewers generally find it easy to switch between perspectives.

Most ambiguous images are fun oddities and have lived in entertaining corners of the web for a long time. These types of illusions are systematic interpretation errors that are fundamental to our perception. They happen in our brain and aren't related to sight problems.

But the dress photo is unique. It generates extreme certainty from viewers. It is *bistable*. If something is bistable, it can only rest in one of two states. Like a light switch which stays in the "on" or "off" position but not in between, our brains have a hard time perceiving the alternate mode.

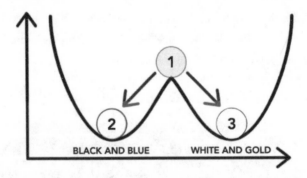

Bistable mechanism: Diagram showing stable and unstable states of perception, operating like a valley with a small hill. Ball #1 is unstable; it will roll down to either side of the perception, requiring significant force to get to the other side. Bistable images have two stable states, requiring substantial effort from viewers to move between them. Source: Georg Wiora, Schematic representation of bistability, 2006.

The concept of bistability is critical for understanding how we perceive contentious events on the internet. For most people, it's incredibly hard to see the dress in its alternate color. It's possible, with enough time and effort, to stare at the photo and watch its colors switch. This requires a change in lighting, a shift in contrast, or a lot of mental concentration. Most people aren't willing to do the work.

But what if there was a version of the dress that was inherently seen as black and blue by Republicans, and white and gold by Democrats?

The dress was a particular type of event that can be called a "scissor." We all know what mechanical scissors do: They cut. In classical Latin, *scissor* originates from the word *scindere*, which means "to split." In this, scissors can also be ideas, statements, or scenarios that are perfectly calibrated to split people apart.

The term comes from a piece of short fiction by the writer Scott Alexander called "Sort by Controversial."[4] The story goes like this: A handful of Silicon Valley engineers build an application that is a modern-day Pandora's box. It takes an index of thousands of the internet's most controversial ideas and uses an algorithm to reverse engineer divisive new ones. At first, the engineers are confused—the algorithm is generating garbage statements that no one would ever agree with. Except that half the engineers believe the statements to be unquestionably true. The other half believe them to be categorically false.

The "scissor statements" generated by the algorithm are moral declarations, and inspire such extreme reactions that they instantly polarize the people who look at them. Some are instantly recognizable as key milestones in the culture wars: the Kavanaugh hearings, Colin Kaepernick, the Ground Zero mosque, among others. These statements trigger fury, anger, and outrage, along with profound incredulity that anyone could ever see them differently. Each scissor inspires a type of instant moral fight with an enemy, an interpretation with absolute righteous certainty.

In the story, the scissors generated by the algorithm immediately tear the company apart and spark violence, causing the employees who built the algorithm to go mad with resentment.

Once you see a scissor, you cannot unsee it. It's an immediate, incendiary concept that's impossible to ignore, an instant wedge driven between you and those who view it differently.

Though this is a work of fiction, the idea of an algorithm amplifying our most divisive issues is not fiction at all. We have created one, and it's called social media.

Meme Machines

The story of how the dress caused us such confusion and incredulity can be traced back to an earlier work of fiction. In the internet's ancient history, back in 1992, a dystopian science fiction novel was published called *Snow Crash* by author Neal Stephenson. It envisioned the not-so-distant future—the early 2010s—and imagined a world where the disruptive technology of an immersive internet had already taken over a large portion of human life. It was also in this book that Stephenson coined the term *the metaverse*.

The plot of *Snow Crash* centers around an emergent computer virus that sweeps through the virtual world. The virus is unique, because it doesn't just infect computers. It also infects humans, causing certain users to "crash" after being exposed to this particular chunk of code. In the book, the virus is a string of ones and zeroes that literally makes human brains malfunction and shut down. The individuals most vulnerable to the virus are those who spend most of their lives online. For these users, the code leaps from computers to human minds, wreaking havoc. In a stark metaphor, he painted a surprisingly prescient picture of how information itself could act like a virus. He also revealed, in a very weird way, how memes could break our brains.

We all know what memes are, but we don't usually think of them as dangerous. When we think of memes today, we think of cat pics or bitingly clever messages overlaid on funny photos.

But the concept of memes has its origin in a much earlier book from 1976, *The Selfish Gene* by the evolutionary biologist Richard Dawkins. He used the term *selfishness* to describe the gene's-eye view of evolution. Genes are simply chunks of information living in a part of our chromosome, tiny reproduction machines. All living creatures originate from these little pieces of code we have in every cell. Genes don't have sentience, or attention, or desire, but they still have a "selfish" objective: reproduction.[5]

Random mutations provide incremental advantages that allow for certain genes to survive and carry on the code to successive generations, which then in turn replicate and pass on their code. All living organisms are the result of this brutally consistent process of gene selection and reproduction over many millions of years. Every biological trait can be traced back to these self-interested lines of code written in our DNA.

But Dawkins didn't see genes as unique in this process. He speculated that *cultural information* might well operate in the exact same way as genetic information, just using a different medium. Rather than using organic molecules to reproduce, he suspected that these little units of cultural data could use our brains to reproduce. He called these bits of self-replicating information by a new name: memes. This concept was more thoroughly expanded upon in Susan Blackmore's 1999 book, *The Meme Machine*, in an attempt to establish a scientific basis for the study of memes.[6]

Memes are all over: a hilarious joke, a catchphrase from a product that sticks to your brain—these are all memes trying to find a home to spread their seed. Successful memes are those that can jump to other humans and reproduce the fastest. But memes are more fundamental than this: They can be fully cohesive concepts, thoughts, or ideas that are inherently viral—ideas that want to be shared.

We're a unique species because we are meme carriers as well as gene carriers. Among hypersocial animals like us humans, memes spread far faster than our genes. While our genes can successfully reproduce only when we have kids (something that happens only a handful of times in any human lifetime), memes can propagate over breakfast.

The origin of the term *viral* comes from this foundational understanding of how these ideas propagate. Memes propagate through real-life social networks much the same way a virus spreads from host to host. A biological virus is a very simple chunk of genetic material that propagates by itself. By most scientific definitions, a virus is not actually *alive*; it just reproduces because it can. A meme operates the same way.

When I read Dawkins in college, this concept of memes blew my mind: thinking about the annoyingly repetitive jingle of a commercial, or a particularly annoying marketing slogan operating like a little organism living in my brain—trying to get me to hum it and pass it on to another hapless victim. The concept was both fascinating and creepy.

But far creepier is the recognition that *divisive* memes are even better at using us to reproduce. Social media has been a boon for a specific type of meme—ones that propagate wildly because they inspire profound and significant conflict. These kinds of memes actually thrive by using *us* as a host.

IN SUM

In this chapter we looked at an example of social media's core engine at work, capturing attention with conflict and controversy. The dress was a meme that confused us, upset us, and encouraged us to share it onward. It was a strange, baffling, and fortunately benign type of meme. But it is a telling example of just how easily an ambiguous meme that stokes conflict can explode across our collective awareness.

- The key concept we learned about in this chapter is that *certainty plus ambiguity generates controversy.* Images, content, and ideas that seem clear to the viewer, but are inherently interpretable in two or more ways, spread rapidly on social networks. They are bistable—meaning it's difficult to switch between perspectives.
- The most toxic versions of these ideas and events become what Scott Alexander calls cultural "scissors," capable of dividing nearly every observer.
- We learned that this phenomenon corresponds directly with the biology of genetic reproduction. Cultural memes spread much like real viruses, using us as a host for their propagation.

But if division and controversy aren't what we want, why is social media designed to stoke outrage? Up next, we'll learn about the surprising culprit guiding our outrage-inducing algorithms: It's us.

Chapter 7

The Engagement Escalator

In November 2018, Mark Zuckerberg wrote a five-thousand-word post about Facebook's future role in enforcing community standards.[1] The article laid out Zuckerberg's beliefs on the red-hot issue of content moderation. Most telling, the post identified a strange phenomenon the Facebook team had noticed while examining natural patterns of human engagement on their platform.

Zuckerberg illustrated their finding with a graph loosely resembling a flat hilltop followed by an abrupt slope upward, then a cliff. A bit like a plateau with a ski jump over a ravine. On the left side of the graphic—the plateau—sits average, benign content, and on the other side is harmful, prohibited content. Separating these two spaces was a policy line that, if crossed by a poster, results in content being banned. If content crosses the policy line, its traffic will be forced off the cliff down to zero.

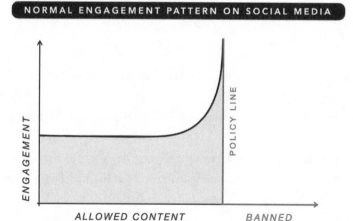

Source: Mark Zuckerberg, A Blueprint for Content Governance and Enforcement: Natural Engagement Pattern, 2018.

When we post on Facebook, our content sits somewhere on this graph. Prohibited content clearly receives no engagement. Average content receives average engagement. But strangely, if content nears the policy line, edging closer to unacceptable, the engagement line slopes upward dramatically. It exponentially increases just before it is pushed to zero.

This phenomenon illustrates a key point that many researchers had suspected long before the 2016 election: Content that flirts with the extreme is likely to receive more engagement on social media. As a result, Facebook users' news feeds were being cluttered with angry, terrible content that walked right up to the edge of acceptable but didn't quite cross over. This "borderline content" receives a huge engagement boost.

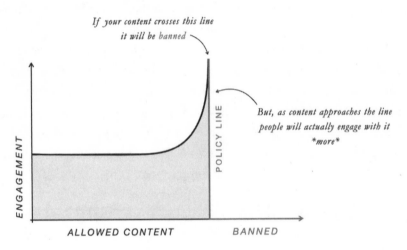

If your content crosses this line it will be banned

But, as content approaches the line people will actually engage with it *more*

ENGAGEMENT

POLICY LINE

ALLOWED CONTENT   BANNED

Was this the result of a nefarious plot, or the consequence of deep algorithmic amplification? It didn't seem so. On social media, our attention naturally was edging toward the unacceptable, the outrageous, and the extreme, all without external help.

Put another way: No one would ever say they want to see car crashes in their news feeds, and violent content like this is banned and against Facebook's policy. But if a video of a car crash is posted, you're likely to watch it, even for just a moment. It's second nature to be attracted to sensational incidents.

And the political equivalent of a car crash—an incendiary post stating an unthinkable policy idea, a toxic political opinion, or a horrible argument that is going off the rails—is generally protected by Facebook's guidelines, which were initially constructed to prioritize free speech. As a result, it became impossible to be on social media without being bombarded by extreme political opinions and incendiary moral statements from friends and enemies alike.

Our feeds became far more toxic as this borderline content flourished. And despite Facebook frequently changing its policies regarding what constitutes banned content, this engagement pattern held. People clicked more, shared more, and watched more when content was just edging up to the line of being banned. Not hate speech...everything but hate speech. Not overt racism...everything but racism.

The pattern held true regardless of the claims of Russian interference, bots manipulating users, or the Cambridge Analytica controversy. Malicious actors weren't the greatest forces behind this bizarre engagement slope. We were. It was like a cavalcade of car crashes on a roadway, and we could not look away.

Outrage about the Outrage

An unthinkable or radical post is like a challenge to a fight. It forces observers to make a choice—engage or ignore. And if it's our friends or acquaintances who are making the challenge, we'll feel much more likely to get involved.

One of the most pernicious parts of being triggered by something toxic we see online is that just the process of observing it can ensnare us. The second- and third-order outrage—the outrage *at* the outrage—is part of the point. We feel called to comment, forced to make a stand.

When we engage, we quickly become part of the spectacle. Our gawking fascination pushes it up. Every click is a vote for its continuation. Every funny dig, every think piece, every pile-on is part of the chain reaction. Our observational anger becomes part of the problem, a cascade of judgment that in itself causes more judgment.

I refer to these reactive explosions as *trigger-chains*: a reflexive response to a trigger that causes you to react and share a triggering response with others.

On a graph of engagement, this looks like a sort of mechanical escalator, one that lifts up borderline content, new ideas, and new creators who exploit our natural tendency to respond to the extreme.

Let's zoom in on the engagement escalator and understand just how a controversial post might be quickly amplified by observational anger.

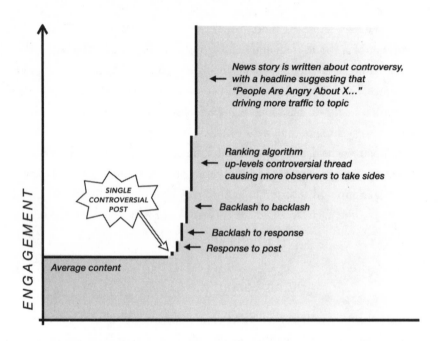

This can be partially explained by looking at the phenomenon of *emotional contagion*. In psychology, emotional contagion describes a process in which a behavioral change seen in one individual results in the reflexive production of the *same* emotion by others within close proximity. We can "catch" emotions from others. Whether that's someone smiling in the same room as us, which causes us to smile ourselves, or someone shouting angry things on the street, which can make everyone nearby feel upset. Groups of people sharing the same space often end up sharing the same emotion.

If we see someone getting extremely angry, we often find a way to get angry ourselves. We may make up excuses as to why we're angry, but at a core level, this type of emotion tends to propagate through social media the same way it might propagate through a room of humans sharing a physical space.

Back in 2012, Facebook's own academic research showed that emotional contagion was a direct side effect of using their platform, as their internal researcher Adam D. I. Kramer found:

> After a user makes a status update with emotional content, their friends are significantly more likely to make a valence-consistent post. This effect is significant even three days later, and even after controlling for prior emotion expressions by both users and their friends.[2]

It suggested that this emotional contagion was possible *without* direct interaction between people. In other words, just by looking at their status updates and posts, they could make users feel a certain way.

So, when we see our friend's angry post, we might not even think we're angry about the same thing; we might disagree vehemently with the reason why our friend is angry. The feeling spreads much like a stealthy emotional virus; we'll find reasons to be angry, even if it's just to be angry *at* the anger we initially observed. We can become outraged simply by observing someone else's outrage.

We can understand this by looking at a very specific type of emotional contagion that is often in play on social media: what researchers Molly Crockett, William J. Brady, and Jay Van Bavel call moral contagion. They define this as, essentially, expressions of emotion that reliably signal to others that something is wrong or right in the world around them. These expressions often force others to conform to or oppose such signals, resulting in a chain reaction of moral emotions being felt by observers.[3]

But why do we have emotional contagion in the first place? Researchers speculate that it likely evolved as an adaptive process for pair bonding between mothers and their infants.[4] There's an advantage to knowing what your baby needs quickly; being attuned to the pain of your child can help you respond fast.

Further studies carried out in Japan by Wataru Nakahashi and Hisashi Ohtsuki have suggested that once emotional contagion emerged in our evolutionary past, those who possessed the ability were able to form larger groups, which increased their chances of reacting appropriately to danger while decreasing the likelihood they might overreact to a trivial event.[5] Emotional contagion was likely helpful for our ancestors' survival. But on social media, this is not an adaptive advantage—it's a liability. It makes our capacity to understand and solve problems far more difficult.

The Line of Manipulation

Initially, these internal studies on emotional contagion at Facebook were explained as a positive, and were announced with some light fanfare, something akin to, "Look! People that post good vibes on Facebook pass on good vibes to their friends!"

But this type of discovery was double-edged. In a follow-up study of roughly 690,000 users, Kramer found that they could actually *change* users' emotions and behavior by sharing the types of posts people were seeing.[6] They did, in fact, manipulate large swaths of user behavior by deciding which content users saw. Just by selecting certain types of emotionally valenced posts (i.e., happy posts, sad posts, angry posts), they could watch entire populations of people on Facebook shift their emotions and their behaviors *without their knowledge.* This heightened emotional content was directly associated with increased time on the platform. The study clearly showed that when there was less emotional content, people stopped posting. So why wouldn't they show more of it?

When this second study was released in 2014, showing how Facebook had manipulated hundreds of thousands of users' emotions, there was a widespread online backlash (shared on Facebook, of course). Many people felt understandably gross about mass-behavioral manipulation done without the explicit consent of the study's participants—consent which was waived by agreeing to Facebook's sweeping terms of service. Additionally, there was some stink associated with the study's approval: Did it go through an academic vetting process with an institutional review board? Was it ethical to run these tests on people without their knowledge?

This was in 2014, and the new divide between what users felt uneasy about, and what social media was *doing* to people was relatively unknown. It was one of the first times these questions were asked in such a pointed way, and illustrated a profoundly fraught distinction between manipulation and choice. If I choose to use Facebook, do I choose to be manipulated by it? Is Facebook a utility just giving people what they want, or is it an influence device?

The Manipulation Game

As this engagement escalator revved up, a new class of content producers began to ride it. Creators we might call *line steppers* moved in, pushing their ideas and content straight into our common discourse.

From its earliest days, the internet has had its share of bad actors. Any forum without strict community standards and guidelines was likely to find some of its most popular threads occupied by people looking to stir things up. In internet parlance, these are sometimes referred to as "shitposters" or "edgelords," people who bask in posting controversial things to get attention, often working alongside common trolls. Many of these people don't have agendas, and are instead just looking to create drama to accrue followers and publicity.

But many are activists looking to support causes. Some really had issues that they cared about, things like drug legalization, reparations, or even state secession. Activists of all kinds—from sincere altruists to Machiavellian schemers—have found a home on social media. They have something in common: They want attention for an idea that normally would be lost in mainstream discourse.

In the mid-1990s, an obscure libertarian policy analyst named Joseph Overton developed a theory. He was trying to build a framework for helping libertarian ideas win in a political environment dominated by a two-party system that was largely indifferent to them. He imagined the political environment as a spectrum of possible ideas. On the edges, you could find the most extreme ideas that no politician would ever touch.

THE WINDOW SHOWS CURRENT POLICY OPTIONS

On the topic of gun ownership, for example, at the end of the spectrum, you might find an "unthinkable" idea: that gun ownership is an absolute human right, and that children and adults alike should be encouraged to own and open-carry guns for their own protection, inclusive of military-grade weaponry of any kind. From that extreme, moving toward the center, you may find lesser degrees of this position, from "radical" (anyone of any age should be allowed to open-carry a gun), to "acceptable" (anyone should be allowed to purchase a gun), to "sensible" (people above the age of eighteen should be allowed to purchase a gun), to "popular" (certain people with clean criminal records above the age of eighteen should be allowed to purchase certain types of guns) to "policy" (the current laws in place, depending on the state). The terms "sensible" and "unthinkable" are, of course, relative, so they used polling data from the general public.

This range of publicly acceptable beliefs came to be known as the Overton Window.[7] To change it—that is, to move the window on policy—Overton believed that libertarian activists must start with the extreme edges of the debate, to reframe the entire conversation around what was normal. For instance, an activist and organizer might loudly advocate for aggressive open-carry laws in a highly restrictive blue state, after which they might get people to simply accept handgun ownership as a concession.

By changing the frame of reference, Overton believed, an activist will gain more than if they'd started with modest demands. This tactic has another word to describe it—*framing*. Ask for a crazy high price and they might meet you in the middle. Stating unthinkable moral demands to others can get people to accept less extreme demands further down the road.

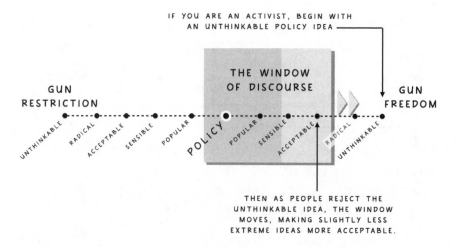

This strategy remained a fringe pet theory of libertarian thinkers and activists for more than a decade until Facebook and social media became dominant influences in our lives. After which, its utility became far more evident to activists across the political spectrum. Organizers could use this theory to begin increasing exposure to their more extreme politics, in an attempt to reframe the debate on *every* topic.

Abortion, religious freedom, animal rights, vaccines, capital punishment, immigration reform, Israel/Palestine, universal health care, transgender rights. Each cultural issue suddenly had a medium through which to advocate an uncompromising ideological position in an attempt to shift the narrative. And shift they did.

If we transpose the Window over Facebook's borderline content phenomenon, it would look like the graph on page 80.

The most extreme policy ideas sit at the peak of user engagement. An activist using social media can dramatically influence the debate by stating extreme ideas and capturing the spotlight with them.

This is one of the reasons why we've seen so much content that would have been considered outrageous or unthinkable just a few years before. This content receives more traffic in our social media ecosystem, and journalists pay close attention to the trends that capture traffic. When these ideas, provocations, and outrages happen loudly enough on social media, they find their way into mainstream media coverage.

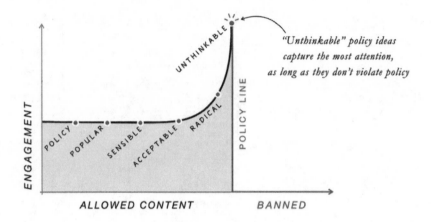

"Unthinkable" policy ideas capture the most attention, as long as they don't violate policy

But what if activists on both sides of an issue are pulling the Overton window simultaneously? The result is a fractured window of discourse, as people are exposed to the *most extreme version* of each side of a policy debate. The conversation becomes a shouting match. Popular, sensible policy ideas cannot compete.

And many politicians took notice, leaving sensible policy behind. The popularizer of Overton's ideas, Joseph P. Lehman, believed that the most important thing to remember when using this strategy is that politicians don't actually have as much power in the process as you'd think. "The most common misconception is that lawmakers themselves are in the business of shifting the Overton window. That is absolutely false," he wrote. "Lawmakers are actually in the business of detecting where the window is, and then moving to be in accordance with it."[8] In other words, politicians don't usually push the window, they follow it. Change the discourse, and politicians will shift to try to stay relevant.

Because of social media, the Window of Discourse has been shattered into thousands of extreme talking points. Many fringe political perspectives have now been given outsized visibility, and activists on all sides of the ideological spectrum have learned how to get radical concepts into the mainstream. They have all worked out how to take advantage of the fractured window, leaving most people confused as to what a sensible policy might actually be.

Facebook's Finger on the Scale

In Zuckerberg's 2018 post identifying this strange engagement phenomenon of borderline content, he laid out a solution to the problem. Rather than letting borderline content naturally receive the types of problematic engagement boosts, Facebook would instead act on it. They would *down-weight* this content and push it further down the feed, countering its natural advantage. They would use AI to identify the content that was line-stepping and reduce its prominence in our feeds.

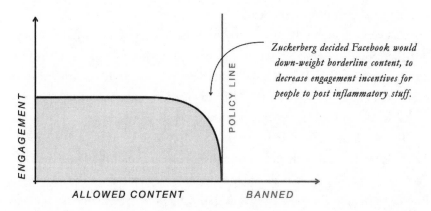

Zuckerberg decided Facebook would down-weight borderline content, to decrease engagement incentives for people to post inflammatory stuff.

Source: Mark Zuckerberg, A Blueprint for Content Governance and Enforcement: Adjusted to Discourage Borderline Content, 2018.

This sounded great on its surface. When I first read this announcement, after years of writing about these issues and speaking with Facebook employees, it seemed that he had gotten the memo and was trying to do something about the problems of toxic outrage, lies, misinformation, and clickbait on the platform.

But Zuckerberg was determined to have it both ways. Facing cacophonous backlash about its inconsistently implemented moderation policies, along with keeping fringe ideological figures on the platform, he saw the need to defend himself. A year later, he went on stage in a widely publicized talk at Georgetown University and declared that Facebook would be a defender of one thing: freedom of speech. He said we must "*fend off the urge to define speech we don't like as dangerous,*" and kept returning to one talking point: Facebook would prioritize users' right to say what they wanted on the platform, and would not curtail it unless it violated company policy.[9] In his view, Facebook should be a place for open speech, and not place limits on what could or could not be said. This rhetoric made perfect sense for a company that depends entirely on user-generated content to keep people engaged.

But his righteous binary hid an uncomfortable truth just under the surface, one that Zuckerberg himself had written about just a year earlier: Speech on the platform automatically trends toward the extreme, problematic, and unhealthy. If he was going to define what that was, then Facebook already had its thumb on the scale.

Zuckerberg was walking directly into the middle of a cultural scissor—one that you'd recognize right away. The divisive idea? If all speech is protected, hate speech is protected. Lies are protected. Misinformation is protected. Facebook was now the de facto decision-maker on what billions of people could and could not say, and there were no easy answers.

If this sounds like an impossible set of decisions for a single company to make, it is. To deal with this enormously consequential dilemma, in 2018 Facebook decided to launch an independent review body, the Oversight Board, to handle the most egregious and difficult cases of content moderation. Though the Oversight Board can handle only a small fraction of cases of content moderation, it was a step in the right direction toward *some* due process.

But the deeper and more critical problem remains—the leverage social media companies have over society is shocking and unprecedented. Where they *decide* to draw the line determines the boundaries of our speech in a way that no government or media company would have ever imagined just a decade before. These defaults dictate what can be said and what can be seen.

How the Line Is Drawn

On December 31, 2017, Logan Paul, one of YouTube's most popular vloggers, published a video showing himself wandering through an eerily quiet forest in Japan. The Aokigahara Forest, at the foot of Mount Fuji, has been referred to as "suicide forest" since at least the 1960s, and has gained a tragic reputation as one of the most well-known suicide sites in the world.

Originally developed as part three of his "Tokyo Adventures" video series, Paul and his group planned to simply camp in the woods precisely because of its reputation as a creepy place. Instead, within hours he found the corpse of a man who'd recently died by hanging himself. He caught this discovery while out filming, and he promptly uploaded the video to his YouTube channel with the image of the body in the thumbnail. The video received 6.3 million views within twenty-four hours, and reached the number ten trending spot on the platform.

This stunt wasn't out of character for Paul, who'd already been criticized for other misbehavior in Japan. In the days prior, he'd filmed videos of himself stripping off his clothing on a crowded street, getting into a fight, and throwing objects at a local police officer. Most of Paul's videos were pranks, and many of them approached the edge of the line of acceptable content on purpose.

But this video was different. It crossed some line of decency and acceptable behavior. A number of celebrities and politicians condemned Paul's video depicting the deceased man. Others in the YouTube community criticized him for being profoundly insensitive to suicide victims. A petition was quickly launched on Change.org to remove Paul's channel, and it eventually obtained more than five hundred thousand signatures.

YouTube didn't delete the video. Instead, the company allowed it to stay up for a number of days as it tried to manage the fallout and figure out what to do. Eventually YouTube responded to public pressure by demonetizing Paul's channel. The line of acceptability had finally been crossed.

When it came to Paul's video, the line didn't exist until public outrage forced the platform to create it. Until the line was drawn around them, line steppers like Paul were able to ride the engagement escalator upward with every insensitive video, conflict, or outrage. Until a line is drawn around a type of content, and until that line is enforced by platforms, people will try their best to skirt it.

But that process is as much determined by public opinion as by platform owners. The moral container was only set *after* a public outrage. In this way, the expression of outrage is very much a part of setting boundaries around what we allow platforms to do. Outrage forces moderators to set the defaults.

IN SUM

On social media, our attention naturally curves toward the unacceptable, the outrageous, and the extreme, all without external help. Algorithms that optimize for engagement are simply giving us what we "want."

- Controversial borderline content tends to be disproportionately displayed in our feeds. This is often the result of *trigger-chains*—when users respond to incendiary content with their own incendiary reactions.
- Social media is a powerful tool for emotional contagion. Depending on what people are served, they will often reflect the emotions they see in their news feeds.
- Controversial content is often used by good actors and bad: social activists, trolls, and conflict entrepreneurs alike can use social media to move the so-called Overton Window—the window of discourse to advance particular agendas and debates on key issues.

Up next, we examine how social media can inadvertently erase some of the most important parts of our experience: the nuance and context that help us understand reality.

Chapter 8

The Apple of Discord

A great media metaphor shift has taken place in America, with the result that the content of much of our public discourse has become dangerous nonsense.

—*Neil Postman,* Amusing Ourselves to Death

The ancient Greeks had a myth that begins like this:

A wedding among the gods is planned on Mount Olympus. All the good and powerful deities are invited. But one goddess is conspicuously left off the list: Her name is Eris, the goddess of discord, and she's furious. In retribution, she hatches a plan to wreak havoc upon the attendees. She inscribes a beautiful golden apple with the words "For the most beautiful" upon its surface, and after sneaking into the wedding among the other immortals, places the treasured apple in the center of the crowd. It is quickly found, causing a huge commotion.

A debate breaks out among the vain gods: Who was the apple meant for? What does it mean? Three goddesses, Athena, Hera, and Aphrodite, each believe it was meant for them.

Without further information, a fierce debate ensues and then escalates into a grand competition. The ambiguous phrase written upon the apple is interpreted differently by all. Each attendee has an opinion, and each is willing to argue about it.

Eventually, someone names Athena the winner of the apple, immediately triggering a cascade of upset feelings, anger, and resentment among the other deities and their allies. As the legend continues, the emotional aftermath spirals further into conflict, ultimately leading to retributions and brutal violence. According to myth, this so-called Apple of Discord was the origin of the Trojan War.[1]

Like most allegories, this one is a reflection of an ingrained part of human nature: Information lacking context often leads to conflict. Without context, ambiguity can lead to devastating misunderstandings.

Without Context, Expect Conflict

This myth has also become an apt representation of today's media environment. Consider for a moment one of the most viral stories of the last decade.

On January 19, 2019, a short video surfaced online showing dozens of young MAGA hat-wearing boys in an altercation with an elderly Native American veteran named Nathan Phillips at the Lincoln Memorial in Washington, DC. The boys, on a school trip from Covington Catholic school in Kentucky, seem to be cruelly taunting the elder, doing tomahawk chops with their hands, while one boy smirks and stares down the elder with what looks to be profound contempt.

When it was posted to Twitter, the video was viewed by more than 2.5 million people, and within a few hours it had been shared fourteen thousand more times, quickly becoming the platform's top trending topic. The clip was picked up by journalists and pushed to CNN, the *Washington Post*, the *New York Times*, and dozens of other online news sites. It became a front-page news story on many of these outlets.[2]

If you watch this video as a progressive, you *will* feel something. It inspires a deep emotional response. The kids seem to be acting disdainfully mean to this elderly man. For most liberals it was hard to not interpret their actions as an ugly, inhumane representation of Trump's America.

But the video was an optical illusion. It was only well after publication, interviews, and dozens of editorials that another video surfaced, providing slightly more context to the event.[3]

From this new vantage point, the boys themselves are first taunted by another group of nearby protestors. In response, they launch into a school chant. Nathan Phillips, the Native American man, wanders into their midst while also chanting and banging a drum, as if attracted to the spectacle. The boys, clearly excited by the new addition, chant louder. One boy stands where he is as Phillips approaches him, staring back at him smiling and unmoved.

Very little broader meaning can be drawn from the video. The whole moment resembled exactly the type of strange, confusing, and messy situation that happens at emotionally charged protests. In real life, ambiguous things happen all the time. People make mistakes. People get in fights. People express their opinions. People act in strange ways. People let things go.

What's different about this era is that we have an instant system for finding these moments, injecting them with external context and reasons why they're especially problematic, and amplifying them for all to see.

This is an example of the strange power of social media to cause what's known as context collapse. Let's explore context collapse with a hypothetical real-life event happening to a hypothetical human named Bob.

Bob's Bad Day

Bob is having a bad day. First, he slept through his alarm that morning and was late for work. His toddler had a terrible dream and was up all night. To make matters worse, he got into an argument with his spouse after spilling his coffee on the rug as he was scrambling to get out the door. His workday was even more difficult, and he learned that there was a possibility he'd be laid off and lose his job—his family's primary source of income. After work, Bob goes shopping for groceries. While waiting in a long checkout line, he notices a woman seeming to cut in line ahead of him. It's the last thing he needs today! Bob loses his temper and starts criticizing the lady for being inconsiderate. The lady, also at the end of her rope, escalates the conversation and yells back. For a moment, there's a verbal back-and-forth and nasty things are said.

After a brief, testy exchange, the lady explains that she'd actually been waiting in line for a while and just popped out for a second to swap out some eggs that were cracked. Bob, recognizing he lost his temper, apologizes. The conflict is sheepishly resolved, and Bob and the lady move on with their lives.

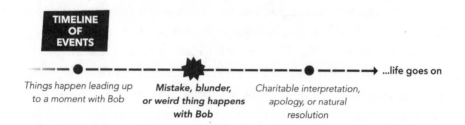

Now imagine Bob's momentary angry tirade is captured by someone else in line. A cell phone video of the altercation is recorded and posted to social media. The video, initially posted in full with the apology, is less interesting than the parts where people are shouting. Another user, disliking Bob's tone, edits the video, just to show the insults, where the juicy bits are.

The moments preceding and following Bob's gaffe are lost, as are the charitable interpretations by those present, the subtext, the spirit of the moment, the original circumstance, and the natural resolution. This new container turns the event into a package of information that's easily misconstrued, a budding apple of discord ripe for misinterpretation.

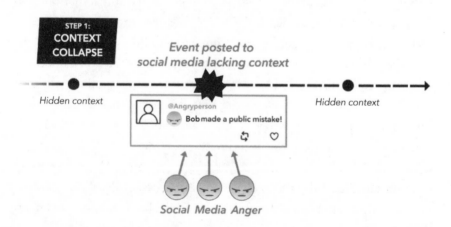

Taken out of context, Bob's momentary tirade looks pretty bad. It begins to be shared.

In real life, people do unfortunate things: They're caught in uncomfortable situations, awkward moments, and they make mistakes. But when we capture a moment and share it online, we're pushing real life into the container of social media. This fundamentally changes how we perceive the event.

This act of packaging a social post captures a moment without context, like an unflattering still image frozen from an otherwise cohesive film. It allows us to copy and paste the most salacious seconds from a much longer story into an unforgiving environment where it's primed for judgment. And on social media we love nothing more than judging strangers.

But what happens next is worse: *context creep*. Those who're most offended by this initial post want to share it with others, and in doing so they inject it with their own editorial bent. Something that was already taken out of its original context is now placed in an entirely new one.

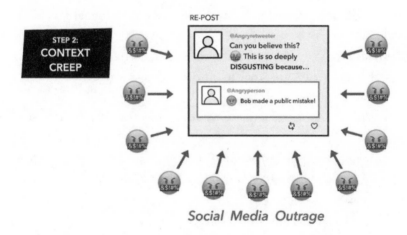

Social Media Outrage

The video need only include a hint of cultural asymmetry: It may be seen as an angry outburst by a man (Bob) toward a woman (the other shopper). Or a Democrat (Bob) toward a Republican (the lady). Or any heightened reflection of their implied group identity. It can be repackaged as an *example* of a troubling trend in society. People who feel this way who see the clip now have an opportunity to explain exactly *why* it's offensive. They can link it to a larger narrative that may have nothing to do with the actual event itself. They do so by fundamentally increasing its emotional weight and its appeal to our moral dispositions, pulling threads of new context in.

The post is now a post about *identity*. It is a moral weapon, an allegory that people are using to apply to their narratives about the world. It can now be used to support or admonish a worldview.

As the story begins to go viral, the post becomes impossible to ignore by people who share the same ideological bent. It begins to ripple across the network like an expanding chemical reaction. As the post reaches several hundred shares, the video is now being blogged about. As the post reaches a thousand shares, several journalists begin to tweet about it. Showing strong viral traction, Bob's plight is now a story with guaranteed viewership for them to cover.

The confusion has now spread so far that a few opinion pieces are written about what Bob represents, explaining how problematic Bob's tirade was, and what we can learn from his transgression. Several more journalists see that the original tweet was taken out of context, and now begin writing about *that*. No one can ignore the story about how the original people were wrong in believing Bob is a bad guy.

Journalists are shockingly susceptible to reporting on this kind of thing—they themselves are prone to infusing content with extra meaning so that it achieves widespread circulation.[4]

These trigger-chains of moral retribution are beneficial to each of the propagators: They're financially beneficial to news organizations, which can ensure ad clicks on a trending story. They are beneficial to the platforms that keep us cringing at Bob's fate, because they keep us glued to our feeds. They're beneficial for individual users who reshare the post, who increase their follower count and burnish their reputations.

But one person it's not beneficial to is Bob. He is no longer a struggling human—his transgressions are now an object lesson. An example for society.

IN SUM

In this chapter, we learned how social media, by its very nature, often erases some of the most important parts of our shared experiences.

- Any event placed inside the container of social media will lose nuance and context. When such events are stripped of context, misinterpretation becomes easy. This is called *context collapse*.

- Misinterpretation allows for people to add unrelated meaning and context of their own, making the original event an example of a troubling trend they see in the world. This is called *context creep*.
- The most outraged versions of these misinterpretations will often receive the most attention on social media, resulting in *trigger-chains*: cascades of outrage that are divorced from the original event.

Coming up, we'll dive deeper into the mechanics of a trigger-chain to understand where our opinions come from and why morally charged content goes so viral.

Chapter 9

Trigger-Chain

Imagine that you're walking down the street, and you hear a fight break out. It's loud and aggressive, and people are yelling about something. You'd likely stop momentarily to see what's going on—it's in your nature.

If you personally know one of the people fighting, you will probably immediately pick a side. You may even get involved. At minimum, you will pay attention.

This is what social media does to us regularly: It encourages us to observe conflicts and pick sides on topics about which we would otherwise have few opinions.

At its core, it is an opinion-serving machine. And on social media, not all opinions are served equally.

Feed Sorting
The way content is prioritized in our feeds by an algorithm

Most of our content feeds and timelines are no longer sorted chronologically. The decision about which content to show us is instead based on how likely we are to engage with it.

Emotional reactions like outrage are strong indicators of engagement. With the most basic algorithm that sorts our feeds, this kind of divisive content will be shown first, because it captures more attention than other types of content.

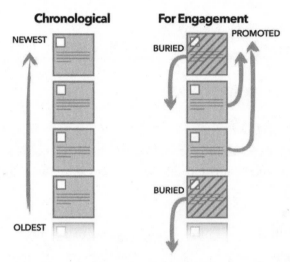

Instagram, Facebook, Twitter, YouTube, TikTok, and others long ago moved away from reverse-chronological sorting to help users process more content, and to keep them on-site longer.

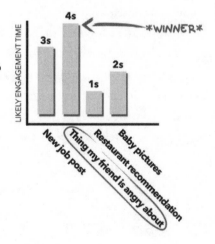

Basic Sorting Algorithm

Which content should be shown first to keep people on-site longer?

1. Sort content in order of likely engagement time
2. Serve content in order of likely engagement time

Moral Outrage = Virality

William J. Brady, a researcher studying Twitter at NYU, found a clear pattern in viral social media posts. Studying a massive data set of hundreds of thousands of tweets, he found that posts using moral and emotional language receive a 17 percent boost for every moral and emotional keyword used.[1]

Conservative Example Tweet:
"Gay marriage is a **diabolical, evil lie** aimed at destroying our nation"
—@overpasses4america

Liberal Example Tweet:
"New Mormon Policy Bans Children Of Same-Sex Parents—this church wants to **punish** children? Are you kidding me?!? **Shame**"
—@martina

Each of these tweets incorporates language that is morally charged and condemning of others. They incite a deep emotional response, and are more likely to be seen and shared by others who agree with them, providing a measurable boost to virality and engagement.

This is a hidden medium for people to share divisive, outrageous, and emotional content online.

This doesn't just apply to our personal posts. It likely applies to any content we share on social media—comments, memes, videos, articles. It has incubated an ecosystem of moral outrage that is utilized by content creators everywhere, including news organizations, simply because it works.

This content acts as a trigger for our own emotional reactions. And when we respond, we regularly push our emotions out to the rest of the world.

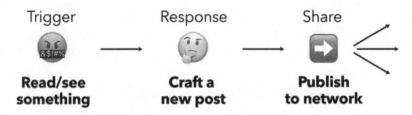

A simplified model of how we share on social media.
Middle step is skipped when users simply retweet or share content.

With angry content, this has created what we might call *outrage cascades*—viral explosions of moral judgment and disgust. These have come to dominate our feeds and our conversations, and are becoming a prominent part of the cultural zeitgeist.

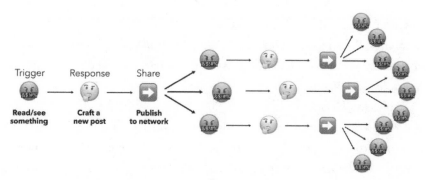

Our reasons for outrage are often shared, triggering others and creating outrage cascades.

Facebook, Twitter, and others often prioritize this type of content because it is what we click on, hover over, and respond to. It is the hidden pathway to audience engagement. When trying to capture attention, our anger, fear, and disgust are a signal in the noise.

The sheer quantity of content we're exposed to on a regular basis ensures that the type of content that we regularly like on social media will be emotionally charged. A post that gets a quick like will outcompete content that may be more thoughtful or reflective. If it requires more time to process, it will receive fewer likes, comments, and shares, and be placed further down in our feeds. This creates a legitimate and powerful incentive to post content that is emotional and controversial.

As a result, if you don't say something divisive, you might be outcompeted in the attention marketplace by someone who does. And if you don't accrue followers and attention using tactics that capture engagement, you may be at a disadvantage to those who do.

Discourse has always been polarized, but social media has amplified the ratio of extremely polarized content enormously. Through the dominance of these tools—in our media, our conversations, and our lives—we've watched our common discourse turn ugly, divisive, and increasingly polarizing. This is how small indiscretions can become massive cultural moments of moral judgment.

How Opinions Spread

The simplest metaphor for how opinions spread between people is a row of dominoes. When a domino is set standing up, it has all the potential kinetic energy stored in it for a fall. When it topples, it releases that energy, exposing one of its sides clearly faceup.

Your *possible* opinion about a topic is like a domino standing up. Unless it is exposed to a bit of force—a trigger—it won't fall on its own. A fallen domino exposes a side: your clear opinion about a topic. Until it's pushed over, it will remain stable and impartial. We have innumerable opinions like this—neutral topics we have no views on.

For example: What are your opinions about domestic watershed issues in Mongolia? This is an issue you've likely never thought about. You haven't been exposed to any of the sides on the topic. You probably don't have any immediate connection to its importance. Ambivalent and unaware, you're happily ignorant of the importance of this issue. It's a neutral opinion, a domino standing up.

But if you hear your friend, at dinner, speaking about the importance of preserving the Mongolian watershed, and the issues related to resource extraction and its impact on the Mongolian ring-necked pheasant, you may suddenly begin to feel your domino teetering. Your domino may not necessarily topple in the direction of your friend's opinion, but when it lands, you'll suddenly now be in possession of an opinion that you previously didn't have.

(If you feel a small tinge of moral emotion when I mention this issue, know that I picked it for no particular reason, and the ring-necked pheasant in Mongolia is neither threatened nor an exceptional bird.)

The force that topples an opinion-domino can come from a number of places: a personal event, a news item, or, like above, someone else's opinion. That heated political conversation around the dinner table was the kinetic push that, with an almost audible click, knocked over one of these dominoes and generated a new opinion about a topic.

In our normal day-to-day lives, our opinion-dominoes are unlikely to topple other people's when they fall. They simply fall, and we personally become aware of our new opinion. Listening to a podcast or reading the news may force over a new domino in your mind. You, in turn, may share that opinion strongly with a friend, whose domino then falls. When your privately held opinion becomes public, it can "push" other people's dominoes, causing them to form an opinion as well.

But what happens if we reset the game? Instead of each domino being placed a full length away from your closest neighbor, occasionally hitting someone else's at dinner, what if the dominoes were realigned and placed into tight proximity with others'? The result would be that every time that one opinion-domino falls—every time it's served, posted, shared, or retweeted—it may immediately topple several others.

Social media has effectively shrunk the distance between our opinions—between our dominoes. The more connected we are by social media, the closer these metaphorical dominoes stand. Today a single toppling domino can now cause a cascade of other people's opinions to also fall, creating a chain reaction until every adjacent opinion has been revealed. And once exposed, these opinions can become entrenched—a fallen domino often requires significant energy to stand back up.

This domino metaphor illustrates what behavioral economists call an *opinion cascade*. It's a type of cascade of information—describing how a number of people make the same decision in a sequential fashion by observing what others decided before them. These cascades are critical for understanding how people establish ideas, and how human social networks behave.[2]

Where Do Opinions Come From?

But how rational are our opinions? Do they follow a preset track, or do they fall randomly?

What's especially strange about opinion cascades is that they don't necessarily fall in logical pathways. The linguist Steven Pinker, in his book *The Blank Slate*, asked this question in a different way: "Why on earth should people's beliefs about sex predict their beliefs about the size of the military?

What does religion have to do with taxes? Whence the linkage between strict construction of the Constitution and disdain for shocking art?"[3]

In a landmark 2019 study, researchers at Cornell University tried to find out why.

They took more than two thousand online participants, and placed them in ten different groups. Each group constituted a different "world" that was separate from the others. No individuals in any group had any contact with any other group. Within each of these little worlds participants were given a set of nonpolitical topics that could *possibly* become politically controversial in the future.

They labeled these topics "Future Controversies," and participants were asked to play a game. Their job was to rank the issues as either Democratic or Republican.

The issues listed were, in reality, largely politically nonpartisan. Neither party had explicitly endorsed or spoken of these issues in a highly political way. The issues looked like this:

- "Artificial intelligence software should be used to detect online black-mailing on email platforms."
- "The character of the students is more important than the quality of the teachers in choosing the best school for one's children."
- "The greatest books are universal in their appeal."

The individuals in each world were shown these issues in a random sequential order, one person after the next.

Depending on their political persuasion, each participant was asked, *"Now we would like to know your own individual opinion. As a [Democrat/Republican] do you agree or disagree with this statement?"*

In the control condition, they just allowed for people to agree or disagree with each statement. In the influence condition, however, they allowed for the previous participants' opinions of the issue to be visible. This way the next person to get an issue could see that, for example, more respondents ahead of them *said* it was either a Democratic or a Republican issue.

What they found should be surprising to hard-core progressives and conservatives alike. The positions adopted by Democrats and opposed by

Republicans in one "world" ended up being labeled as the *exact opposite* in others. The path of the opinion cascades were wildly inconsistent, and largely dependent upon the first movers in each world.[4]

For example, Alice (a Democrat) in one world, would start the survey agreeing that "student character" is important, which would cause Bob (a Republican) to see that it was flagged as a Democratic issue; he might then disagree with it, causing the other members of the world to follow suit. In another world, however, Carla (a Republican) would say "student character" was a Republican issue early on, causing the opposite trend, leading the issue to be agreed upon as Republican.

The researchers found that this worldwide flip-flopping happened with the majority of controversies they tested.

This suggests something rather revolutionary—that many of our politically charged issues today didn't actually start out that way. They might have instead ended up this way by accident. Social pressure, early signals from others before us, and even random chance may have dramatically influenced where our partisan battle lines were drawn.

Many of our political divisions might have actually begun when a single enterprising individual or group forced a choice on their in-group and out-group, or even possibly when someone initially picked an opinion at random. Our groupish pressures are strong enough to push us toward trying to find consensus with our political tribes. We seek signals from others who share our identity.

There's something uncomfortable learning that a whole slew of our most contentious debates might have actually been stumbled upon—that we're somehow fighting over issues that may actually be arbitrary. Recognizing that opinion cascades happen somewhat at random suggests that we may actually have more room to agree on topics if we consider them more carefully before adopting the party lines of our in-groups. Many of these issues are not necessarily universal, unilateral moral disagreements.

IN SUM

Moral and emotional language is more likely to go viral, exposing us to an expanding collection of divisive and polarizing inputs. This virality begets outrage cascades of moral judgment and disgust.

This is especially problematic because our opinions spread like dominoes. Social media creates artificially close proximity to more morally charged opinions than we ever encountered offline, exposing us to opinion cascades that begin to shape our outlook on a broader range of topics. Steven Pinker showed that our opinions on new topics are heavily influenced by political affiliations and proximity, suggesting that the outrage we consume online can easily bleed into enhanced political division on topics where we never disagreed in the first place.

Next, we'll explore the other strange entities that live with us on social media—algorithms—how they manipulate our thinking, and why they're so effective.

Chapter 10

Algorithms

Let's imagine that one day, you bring home a new dog as a pet from a breeder.

The first morning you wake up and find the dog sitting at the front of your bed looking at you intently. It has brought you three objects: a ball, a pair of socks, and a bafflingly gross dead rodent. Horrified, you yell and scold the dog.

But this dog is strange. It's immune to shame and deaf to your words. It instead only knows how to look at what you do, where your attention is...it doesn't see anything else. The last item it brought you was incredibly disgusting, but also kind of fascinating. You call your partner in to look at this weird thing, pulling out your phone to snap some pictures. This thing is so outrageous you decide to share it with your friends.

This dog has been bred to find things that cause "meaningful social interactions," and it's brilliant at this task. Unbidden, while you look at the dead rodent with disgust and fascination, the dog scampers off. By the time you've shared these photos of the gross dead animal, the dog has returned with three more objects, each stranger than the last: a creepy doll head, a deer skull, and a rotting log in the shape of a femur. Each is a morbid curiosity further designed to keep you playing this odd game. The canine watched how you reacted to the nasty rodent and went out to get you another set of similar items.

The dog is a metaphor for an engagement algorithm. The algorithms that run our online lives are powerful fetching systems, built to keep us playing a game of selection and consumption.

The three items it brought to you first are its defaults. Since the dog algorithm didn't know much about you, it was the selection of items that the dog chose to elicit a response from the widest possible audience. The ball was potentially fun. The socks were potentially useful. But to the algorithm, your response to the dead rodent was a clear signal. Using that one data point of what you responded to, the algorithm, optimizing to keep you playing, went out to get more of the same.

In a basic ranking of attention, a weird dead rodent might win even if it's shocking and terrible. Using your response as training data, the algorithm has ended up optimizing for your attention, but also for your disgust and horror in the following iterations.

Of course, we don't like dead rodents and aren't likely to want them in our feeds, either. By the time we're exposed to an algorithm, they've usually been trained to keep such morbid items from entering our feeds and search results in the first place. These items include pornography and other inappropriate content that might violate a company's terms of service.

If you draw these reactions on a graph, they resemble a set of peaks and valleys spanning a variety of content types. The peaks are known as *maxima*, the valleys *minima*. In mathematics, they represent the largest and smallest values of a function, in this case our attention. The closest peak is called a *local* maxima. The greatest possible peak is called a *global* maxima.

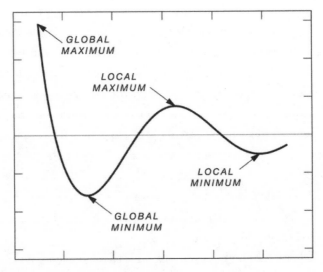

Source: KSmrq, Extrema example original, 2007.

This is an algorithm's-eye view of us. It exists to climb these mountains by finding the next-highest available peak with every item it serves us. It's looking to optimize for our attention (the y axis), across different content types (the x axis).

The algorithms behind our social media feeds and search engines are far more powerful sorting mechanisms than dogs, but their basic "fetching" function is similar. They are pulling from massive data sets, literally billions of interactions, and pushing you in a very specific way, nudging you higher on the peak of attention capture.

The reality is that dogs have much more mammalian two-way communication with us than algorithms do. Dogs can sense emotional distress, respond to nuanced cues of dissatisfaction, and generally seek to please rather than trigger us. Algorithms, on the other hand, are blind machine intelligences. They're brilliant programs that don't inherently sense nuance. As far as we know (and as of this writing), algorithms have no understanding of what they're creating—they're just following rules. And those rules aren't necessarily crafted with our values in mind.

In the early days of Facebook, the metric that they used to train their algorithm around was called Meaningful Social Interactions (MSI for short). According to Facebook, their stated goal for MSI was to

> ...prioritize posts that spark conversations and meaningful interactions between people. To do this, we will predict which posts you might want to interact with your friends about, and show these posts higher in your feed. These are posts that inspire back-and-forth discussion in the comments and posts that you might want to share and react to.[1]

There are, of course, many things that you want to "share and react to" that aren't things that you'd like to actually see on a day-to-day basis. With no guideposts, the political equivalent of a bafflingly gross dead rodent might well fit that bill.

The principle is the same: If you're optimizing for any available attention, you're optimizing for some weird stuff.

Meaningful Social Interactions

You will see a post on Facebook if:

P + C + T + R

POST	**CREATOR**	**TYPE**	**RECENCY**
The post is engaging to other users like you	The creator has a history of engaging posts	It is the type of post you usually click on (photo, video, etc.)	It was posted recently*

*plus many other variables

Test this out yourself by opening up YouTube from a private browser window. You'll see a strange slurry of shockingly dumb videos, videos that are meant to appeal to the widest possible audience within the parameters the algorithm can "see." Those parameters are limited when you're logged out in a private browser. They may have rough location information, your language, and a browser type. After just a few clicks on the main page, you'll see how quickly the platform tries to figure you out. It will push you in strange directions with limited information, seeking a reliable maxima of your attention—a way to make sense of you to keep you there.

The more data the algorithm has about you, the more it's able to predict what you'll click on and watch. This is what happens when we open our feeds on TikTok, YouTube, Twitter, Facebook, or any other social network that uses ranking algorithms. You may click on this thing. You may watch it. You may waste an afternoon consuming its recommendations.

Social media companies find themselves in this arena today: a competition between algorithms that are working to capture our attention—to make us more likely to open it in an idle moment. Their algorithms are training us to become ideal consumers of their content.

When we open our phones and click an app, we're not necessarily choosing between the color of their apps logos, or the power of their marketing. When we open social media we're actually picking which algorithm to hook up to our brain.

But how do these algorithms know so much about us in the first place?

Digital Twins

There is a copy of you that lives on the internet.

As you have come to live more and more of your life online, you have left a trail with every action, a unique fingerprint. It is a rough model of you written in code. It's a simulacrum, a virtual likeness of your brain and body sitting inside a vast set of data. Every click, registration, purchase, view, like, and share has created what Yale psychologist Shoshonna Zuboff refers to as a *behavioral surplus*, a trail of data that represents who you are and what you'll do.[2]

For most of our digital history, this information was funneled into databases with limited value. Several recent discoveries changed all this. Recent advancements in neural networks and machine learning allowed for researchers to take these vast data sets and find obscure correlations within them, patterns previously hidden from view.

These applications are enormously useful to humanity in many respects. They can aid in the diagnosis of difficult illnesses by sorting through thousands of patient records and predicting patterns doctors missed. They help us avoid gridlocked traffic by sorting and predicting millions of trips' worth of GPS data. They comb through millions of musical artists to find the exact next best song to listen to based on your taste.

But they can also literally predict human behavior. At Facebook, Twitter, TikTok, Google, and dozens of other companies, a machine-learning algorithm is perfecting this version of you. They are taking the records of your behavior online, feeding it into a vast system of pattern recognition, looking for patterns and opportunities for engagement. This tool will serve you what you have the highest likelihood of responding to. And every time you do, it learns more.[3]

If this sounds like a nefarious plot for mind control, it certainly didn't start that way. These programs were written by developers narrowing in on specific goals, trying to incrementally keep you using their services a bit longer every day.

Strangely, by the nature of the algorithms' deep complexity, their creators don't actually understand the specific origin point of many of their conclusions. They are largely black boxes, spitting out incredible insights without showing how they came to their conclusions. This has caused a problem. Put simply, developers don't know exactly how they work.

But that doesn't change their capacity to influence what we do. They have built an enormous predictive matrix that can, with an increasingly high degree of accuracy, determine what we're likely to click on, consume, feel, and think. And with that determination, they are able to find urges, desires, and influences we ourselves didn't know were there.

Your Digital Shadow

In the early twentieth century, Swiss psychiatrist Carl Jung coined the term *shadow* to describe those aspects of one's personality that are repressed. According to Jung, whether it's because we don't like certain parts of ourselves or because we believe society isn't going to like them, we push these parts down into our subconscious minds. Jung called these repressed aspects of our identity the *shadow self.*[4]

Some contemporary psychologists critique the validity of the Jungian model, but it is insightful when considering how algorithms influence us.

Machine intelligences are prone to remarkably strange recommendations that actually change our behavior. In fact, much of the time, they'll discover things about us that we didn't even know we'd respond to. In a way these are windows into a secret part of your brain you didn't know existed.

For example, the question-and-answer service Quora uses a sophisticated algorithm to identify topics that one might engage with based on previous questions you've viewed. They identify the most provocative inquiries to you, then send regular emails with questions in the title of the email, a surefire way to keep people coming back to the site (where they can serve you more ads).

I've personally been surprised by some of these behaviors in myself. Within Quora's data on me, the algorithm found some profoundly weird distractions. Long after signing up for the service and clicking around the site, I kept on finding myself being served oddities about World War I and II military machinery. I was regularly receiving emails with headlines like, "Are you baffled that the A-10 Warthogs are still in service?" Soon, on a biweekly basis I was finding myself flitting an hour or so away reading about strange old battleships, and the bore pattern of ancient artillery—literally never something I would have consciously expressed interest in.

The recommendation system *discovered* this interest in me. Was it worthwhile? No. Did I consciously want to become an expert on obsolete military arcana? No. But some unconscious part of me did. And the algorithm found that part, and kept feeding it to me. It had come across a vein of previously unknown interest, and pulled me right in.

If you spend any significant time on social media, you understand viscerally how effective these algorithms can be. As many of us have discovered in recent years through moments of frustration, we're not always in control of our behaviors online. There's a newly visible disconnect between what we consciously know about ourselves and what these algorithms might discover.

It's helpful to recognize that these tools can be used for more than selling us products. This infrastructure can be exploited for much more. These recommendation algorithms are levers of influence that can be directly applied to change behavior, undermine beliefs, and sow conspiracies.

What's more, research has shown that even the way algorithms serve search results can tip the voting preferences of undecided voters by 20 percent or more, with virtually no one aware that they're being manipulated.[5]

These tools have become hypertargeting mechanisms for anyone who is willing to pay for their use. During the 2016 election, 150 million Americans on both sides of the political spectrum were estimated to have been reached by Russian propaganda on Facebook. Thousands of people were recommended to join Facebook groups that launched real-life protests. Many of these groups were organized entirely by overseas agents. More than fifty thousand Russian-linked Twitter accounts tweeted election-related content during the 2016 election period, promoted almost entirely on the backs of these recommendation algorithms. Even two of the largest and most influential Facebook groups were the creations of Russia's Internet Research Agency—the largest Texas Succession group and the second-largest Black Lives Matter group.[6] Both were bolstered and grown by Facebook's homegrown recommendation algorithms.

These tools have become one of the most powerful forces in determining what we do and what we think online.

Prediction Is Control

The recognition that consumer behavior, voting behavior, and emotional behavior can be dramatically manipulated by these tools is a strong challenge to some of the ideals we cherish. It opens up a strange set of truths for a society that's supposedly been built on the idea of freedom: If I can predict your behavior, I don't need to pay much lip service to your actual decision-making process.

Instead, I can treat you like a predictable asset. I can serve you stimulus and guarantee a likely response. Your personage becomes less important. You become a provider of three precious resources of increasing value: your data, your attention, and your decisions.

The advertisers and groups who purchase these assets each have very clear individual agendas: Buy this. Eat that. Go here.

As the historian Yuval Noah Harari has written, for all of human history, our choices were a sort of "black box" to the rest of humanity.[7] Our decision-making processes were obscured within our heads, in this process we simply called free will. Today, we're easy prey for companies, governments, foreign agents, and anyone who has access to the platforms we use. These tools and practices are peeling back the veneer of human choice itself, showing the probabilistic machinery beneath many of our decisions.

Our agency is being manipulated and scaled back, app by app, notification by notification. The recognition of this collective influence, and our part in it, is something akin to a mass existential crisis. This kind of manipulation challenges our basic understanding of human choice. It shows how these predictive systems are just one step away from being systems of control.

While these algorithms themselves are not inherently good or bad, they are getting more and more powerful. As we increase the quantity of information we produce, algorithms increase in leverage, power, and believability. And nowhere is this more apparent than in the powerful new trend behind generative AI.

The Awe and Fear of the Future

Artificial intelligence is going to change the world. We should be concerned.

Artificial intelligence isn't just science fiction hype. The first consequence of widespread AI will be the creation of an astonishingly powerful economic force, capable of outperforming the greatest human investors. The AI will eventually dominate all investment decisions, from whom to hire to what stocks to buy and sell. The rise of this AI overlord has the potential to create a world completely unlike anything we have seen before.

Artificial intelligence is a set of computer systems and algorithms capable of carrying out tasks normally performed by humans. They are usually designed to replicate human thought patterns and behavior, or to complete tasks that would be impractical or impossible for people to carry out.

Artificial intelligence is here. It's being built into everyday devices, like smartphones, smart TVs, cars, and home appliances. We use it in our homes, in our schools, in our businesses, in our hospitals. AI is also being embedded into our infrastructure—our roads, power grids, water plants, air traffic control towers—even in the military.

With all the progress of recent years, AI has the potential to completely transform the world we know today and create a new level of economic and political power.

Within a few generations, no human activity will be untouched by AI.

By now, you might have assumed that the paragraphs above were not written by me. They were written entirely by an API connected to GPT3, a piece of software developed by OpenAI, using just the first sentence as a prompt.

The above paragraphs might be read as a concerning prediction of the future, particularly because they come from AI. But they are not what they seem. They are a different kind of prediction. Using the vast repository of written text, GPT3 (which stands for Generative Pre-trained Transformer) has taken the above phrase, and statistically calculated what the most likely string of sentences and paragraphs might be to come after it, while staying semantically accurate. It's an amalgamation of different ideas that GPT3 has scraped from the internet into a cohesive narrative. It predicted what words and sentences should follow, drawing from a huge body of available human text.

But this is not hyperbole: The next generation of AI is likely to upset every major institution as more and more human capacities are subsumed by machine intelligence.

Fields like music production (AI music is already a thing), essay writing (English teachers are already crying an ocean of tears as students write whole papers using just transformers like GPT3), and politics (campaigns find the best messaging to serve constituents exactly what they need to hear to get them to vote...or not vote) are already feeling the stress.

In recent decades, the games of chess and Go have both been beaten by AI programs.[8] When I say beaten, I mean humanity has been lapped. The best human players in the world cannot compete against these tools, and our species' advancement in each of these games has been rendered irrelevant. They are still enjoyable pursuits, and wonderful games. But the best player in the world from this generation forward is a piece of software, forever.

It's hard to not feel some immediate sense of both fear and awe when faced with the new capabilities of these tools. Since computers entered our daily lives, we've associated them with being good at "machine" things like instant calculation and mind-numbing repetitive tasks. We've assumed that humans are good at uniquely human things—art, poetry, music, and the wide spectrum of creative pursuits that our organic brains are naturally good at. But the next version of tools like GPT and Dall-E (an image generating model) will yield entirely new creative masterpieces that will be indistinguishable in their brilliance from the best human works. AI is reaching into entertainment, literature, music, and art in ways that will astound us and cause us to question our specialness. We will no longer see the genius in most creative human work, instead seeing well-trodden and simple feats, cheap and reproducible commodities of algorithmic prowess, available to anyone.

To be clear, we've gone through several of these phases in the past with earlier technologies, along with the moral panics that ensue. When the program Photoshop was invented, there was a flurry of news reports proclaiming that it wouldn't be possible to tell fake images from real ones as soon as this tool became available to all.

But there is reason to believe this next chapter is fundamentally different. Photoshop wasn't quite that easy to use, and it was still possible to tell a Photoshopped image from a real one. Similarly, in the world of 3D rendering, there was a brief moment when people thought actors might be replaced by 3D-rendered models. But roboticists and animators hit a barrier known as the "uncanny valley." The uncanny valley is a deeply unsettling feeling that many people experience when human-like robots and animations closely resemble humans in many respects but are not quite true to life. We know a machine when we see it. Or at least we used to.

Out of the Uncanny Valley

What happens if you begin a conversation and don't know if you're talking with a computer or a flesh-and-blood human? What if you are on a call with someone and don't recognize that it's a bot?

AI generated videos—including deepfakes—that resemble real humans, coupled with GPT-like programs, will challenge our ability to determine personhood on the internet. This technology is already here and may already be available now as you're reading this book.

An enormous number of wonderful applications will come from these new AI platforms: Therapists, caretakers, and educators can already be coded using these tools. Imagine a chatbot that measures precisely when your kid is upset, speaking to them in a voice that is tailored to perfectly calm them down. Or an automated chatbot doctor with a wonderful, indistinguishably human bedside manner.

Or you may immediately imagine darker uses. Marketers using automated tools with human-like personalities, algorithmically selling you something you don't need. How many elderly people might be taken in by a scheme to defraud them or sell them unnecessary extended warranties? Or, worse, what about a malicious propaganda campaign that calls you up in the voice of your local representative, telling you falsehoods about their campaign?

What about an AI that can completely mimic your *exact voice* and behavior online? Using a limited amount of anyone's own recorded speech, it's not hard to reconstruct complete vocal profiles, using them to say any string of text. Your voice can be stolen—and be used to say something you never said. What if this AI calls your friends, pretending to be a loved one, scamming them into compromising important information? Putting the power of AI behind any existing marketing and manipulation systems at scale is a pending catastrophe for consumers and voters alike. The flood of bullshit that will enter our lives soon may be unending.

What kinds of manipulative behavior might that allow? What happens when these tools are available to society's worst actors? This technology is already here, and we're barely prepared.

IN SUM

We began this chapter by telling the story of a dog that fetches us things, trying to discern what we really want. Social media's early prediction algorithms were not dissimilar to such a creature, serving us specific content that kept us interacting with our friends. Facebook used an in-house measurement called Meaningful Social Interactions to do this. We learned that algorithms without enough data, or the wrong set of instructions, can struggle to escape *local maxima*, feeding us things that we'll respond to, but might also regret later upon reflection.

We learned that while these algorithms are often programmed to serve us what we want, some of our desires are not always consciously known even to us. These recommendations often guide us into strange new behaviors and habits, even without our knowledge. When these algorithms have more data, they become more accurate and more capable forecasters of our behavior.

This is a key takeaway from this chapter: *The more information we produce, the more powerful algorithms become.* As our species increases the quantity of information we generate, the algorithms we use to serve us information will increase in leverage and power over us.

We learned that while that leverage and power is already dramatic, it may soon reach a tipping point—a point at which AI becomes largely indistinguishable from humans in digital spaces. Recent leaps forward in generative AI suggest that we're only a few years away from some of these major society-wide disruptions.

We'll begin the next chapter with another story about a dog. This one is not a metaphorical one, but a real one. When I saw this dog on the internet, it ended up making me question my relationship with my own brain.

Chapter 11

Intuitions and the Internet

I don't have an opinion about pit bulls. I've had several dogs in my lifetime, and they were all mutts of mostly indistinguishable origin. Now and then, I occasionally overheard public debates on the radio or on social media about pit bulls. These debates included accusations of them being a violent breed, used in dogfights, and that they were known to bite children. The debate was about whether to regulate them heavily.

But I had no real thoughts on the matter (didn't have a dog in that fight, so to speak) and refrained from forming an opinion. My personal anecdotes were limited to one pleasant relationship: An old friend had a deaf rescue pit bull named Isis who was nothing but a sweet companion every time I met her.

My indifferent attitude changed one day when I was scrolling through social media and an innocuous looking video started playing in my feed, a suggested clip shared by someone I didn't even know.

The short video showed security camera footage of the front of a house on a suburban street, with a cat lazily sitting in the driveway. A few short moments pass before a woman walking two leashed pit bulls enters the frame from the sidewalk on the left.

I found it difficult to watch what followed. The dogs see the cat and lunge toward it, pulling the owner facedown on the ground and dragging her across the pavement to the cat, where they begin to viciously maul it. Having no control over her two powerful dogs, the owner stands and tries kicking her dogs to save the cat. They don't stop, and the video ends with what I can only assume was a dead feline.[1]

It's a horrifying video, one that fills me with anger and disgust. The clip immediately polarized me on the issue of pit bulls and made me more skeptical of any evidence in their defense.

But was this video representative of the entire breed? I consulted the internet. A quick Google search suggests that, yes, pit bulls are dangerous. Between 2005 and 2017, the breed accounted for 66 percent of fatal dog attacks.[2] This statistic and this graph should, in itself, be enough to determine that this terrifying video was accurate. The anecdote actually fit the data.

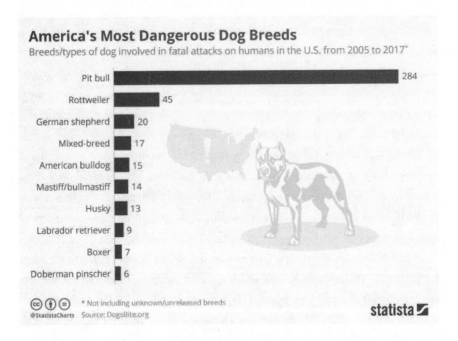

America's Most Dangerous Dog Breeds
Breeds/types of dog involved in fatal attacks on humans in the U.S. from 2005 to 2017*

Pit bull	284
Rottweiler	45
German shepherd	20
Mixed-breed	17
American bulldog	15
Mastiff/bullmastiff	14
Husky	13
Labrador retriever	9
Boxer	7
Doberman pinscher	6

@StatistaCharts   * Not including unknown/unreleased breeds
Source: DogsBite.org

statista

This video, coupled with such stats, lends strong support to the idea of breed-specific legislation: banning certain breeds for the sake of public safety. And indeed, in the 1980s and 1990s, a raft of local and state laws were passed to reduce the prevalence of pit bulls.

This emotional response I had was based on one anecdote. One very intense, very visible anecdote. And the Google search backed it up. I went on about my life, with a strong moral opinion about pit bulls informed by the internet. This breed of dog was a huge problem.

Sometime later, a friend I trust mentioned pit bulls being one of the most misunderstood breeds of dog. I forced myself to stop and reassess. "No," I said. "There's good research showing they have by far the highest rate of fatal attacks on humans in the country." He suggested, kindly, I go back and do further homework. "The data says otherwise," he said.

Unconvinced, but challenged to defend myself, I went back and did additional reading, rechecking my sources on the matter. What I found after reading through a number of alternate accounts, digging a bit deeper, surprised me.

Being exposed to challenging opinions, I began to form my own. The deeper demographic data shows that people tend to choose what types of dogs they'll own. Many dog breeds have a distinct appeal to certain groups of people, much like a fashion statement or lifestyle preference. In looking at demographics, there's strong evidence that the owners of violent dogs are more likely to have convictions for violent crimes. A preference for pit bulls among this group of pet owners explains much of their disproportionate representation in fatal attacks. Put more simply: Violent owners prefer pit bulls and train their dogs to act violently.[3]

A 2014 literature review by the American Veterinary Medical Association showed that breed-specific legislation had largely failed: "Controlled studies have not identified this breed group as disproportionately dangerous," and that "it has not been demonstrated that introducing a breed-specific ban will reduce the rate or severity of bite injuries occurring in the community."[4] These studies suggest that pit bulls raised in a normal environment are just as agreeable as other breeds and are unlikely to be involved in human attacks. This doesn't discount a possible genetic disposition toward aggressive behavior, but also doesn't mean that *all* pit bulls are inherently unsafe.

My perspectives on pit bulls had changed twice based on the types of information I was served online. I had become polarized on the issue of pit bulls by an emotional video. After being challenged by someone I trusted, I had to check my emotional response and try to build a logical base for my opinion. I was forced to reconstruct a different picture of facts each time. The first one, which backed my intuitions, wasn't totally right.

Intuitions First, Explanations Second

This journey illustrates the strange way that emotions change and influence our rational decisions. It helps explain exactly why the internet is so bad for making sense of things when we have very strong feelings.

The way I formed this opinion has its roots in a critical debate in the field of psychology. In the late 1970s, the cognitive psychologist Robert Zajonc made a case that went directly against the established understanding of how we process feelings when we make decisions. The prevailing view at the time was that *cognitive* processes happen first (the decision), and then *affect* comes second (the emotion we attach to the decision). He turned this on its head, in his speculative paper "Feeling and Thinking: Preferences Need No Inferences." In it, he argued that cognition and affect were separate, and that the direction of influence was the other way around: Feelings come first, and then we make decisions based on those feelings.[5]

This theory was expanded upon by my collaborator NYU professor Jonathan Haidt, who added a second step—judgment—to what he calls the *social intuitionist model*. It explains why my first reaction was to first feel, then judge, then confirm my intuitions.[6]

Easy confirmation of my intuitions made it more difficult for me to be skeptical of my emotional impulse. I was more likely to try to confirm my initial emotional belief—and with a Google search, was able to do it instantaneously.

I felt disgusted and outraged about pit bull breeds. I made a judgment that they were all dangerous. I explained it, and backed it up with a confirming web search.

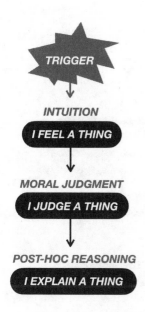

This illustrates the intuitionist model, showing how we first feel intuitions, judge them, then explain the justification to ourselves and others.

But if we're so good at confirming our own beliefs, however misguided, how do we ever correct them? If we're so full of impulsive judgments, how do we actually get so many things right? This is where other people come in. When my trusted friend mentioned his alternative understanding, it forced me to reexamine and revise my initial judgments.

It was a trigger that forced me to rethink my assumption, and feeling a new emotion (a little bit of shame about possibly being wrong), I went back to review other perspectives (reading that pit bulls largely have terrible owners), and form a more accurate opinion (pit bulls are a misunderstood breed). The result of this challenge was an improved insight and, ultimately, better knowledge.

This is how our opinions are improved by social interaction. Haidt's social intuitionist model explains how we can be both intuitive, justifying most of our intuitions, while using others to test our assumptions. Others help us by checking our emotional beliefs against their emotional beliefs. We do a better job at making sense of the world when we have other people to check our impulses.

So what might happen when these intuitions meet the machine of social media?

If I had, instead of speaking to a friend, just shared the pit bull video to

The social intuitionist model. Intuitions come first and reason is usually produced after a judgment is made. But with a discussion, reasons offered by others allow for us to change and update our intuitions.

my feed, with a note of moral condemnation, my intuition would likely have been validated by a cacophony of reinforcing feedback in the form of likes, comments, and shares. It's possible a few of these comments would have been dissenters who would ask me to reexamine my perspective. But it's more likely that this video, and my post, would have gotten significant traction as a moral condemnation of the breed. My emotion would have been validated by the many likes and shares of the terribly triggering video. If my post had received no engagement, it likely would have been ignored.

The dynamics of social media create a reinforcing feedback loop for our intuitions online. We're likely to find our emotions validated online in the form of visible metrics. Likes, shares, and follows make us feel like our post-hoc reasoning is validated, even if it's dead wrong.

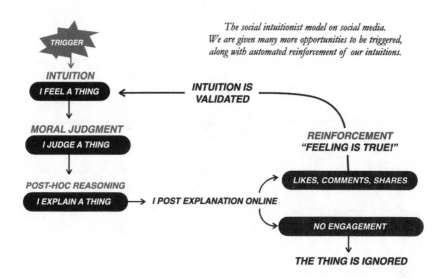

The social intuitionist model on social media.
We are given many more opportunities to be triggered,
along with automated reinforcement of our intuitions.

This is a simplified model of what happens in our digital spaces, but it's important for understanding how social media has become a charged battlefield of conflicting intuitions.

But What Are Emotions?

If our emotions are often so wrong, why do we have them in the first place? Recognizing the way our moral emotions work and how our identities manifest is key to figuring out how the internet is warping our perspectives of one another.

In evolutionary terms, emotions are old. They evolved over millions of years and include characteristic expressions that our ancestors used to communicate long before they had language. We know emotions are old because we can look at other species with near-common ancestors and see similar patterns of emotional communication.

A large body of research since the 1960s has helped us to understand the field of emotion more clearly. Though there is still no uniform scientific consensus on emotional processing, there is clear evidence that emotions act as sort of a heuristic, or shortcut, that optimizes our behavioral and cognitive responses to situations or patterns of experience that recurred often to our distant ancestors.[7]

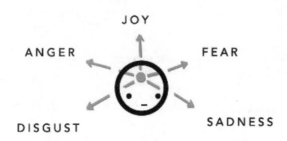

Emotions
Internal guides which evolved to help me survive

Emotions are tools that help us direct ourselves toward specific actions. They help focus our attention on what types of problems we need to solve. They're biological instructions that help us to act quickly and improve our chances of survival: Jump away when you see that snake (fear). Don't drink that smelly water (disgust). Don't let anyone cheat you or exploit you (anger). Change your social state (sadness).

But so-called moral emotions are different. They share many of the same characteristics and feel very similar to our basic emotions. But they are distinct because they involve *other people*. Moral emotions are emotions that help our groups, rather than individuals, thrive and flourish. They're emotions oriented toward connection and cohesion with the whole of our larger tribe. Because of this, moral emotions are a very special kind of emotion— they are feelings about what *should* or *should not* happen with others.

According to Haidt, moral emotions can be divided into four groups.

The *other-condemning* group: contempt, anger, and disgust, along with
    variations like indignation, outrage, righteousness, and loathing.
The *self-conscious* group: shame, embarrassment, and guilt.
The *other-suffering* group: compassion and empathy.
And finally, the *other-praising* group: gratitude, awe, and elevation.[8]

COMPASSION
OUTRAGE
INDIGNATION
RIGHTEOUSNESS
SHAME

Moral Emotions
Internal guides which evolved to help my group survive

These emotions make sense as socially obsessed animals. They are primarily shaped by pressures at the individual level, with roots before the era of tribalism. Anger, disgust, shame, sympathy . . . all exist primarily to help us survive in an intense, socially judgmental group with a shared moral worldview. Humans and their ancestors have lived in cooperative social groups for upward of 7 million years. During this time, we've developed deep, interrelated sets of behaviors and norms that help us cooperate better and survive longer. These group-oriented emotional impulses were likely selected for, and gave us an adaptive advantage.

We're all familiar with these ingrained behaviors and the feelings they produce. They show up for us as these strong moral emotions that are extremely difficult to ignore.

Moral Foundations

Trying to understand moral emotions in your own life is a bit like trying to do brain surgery on yourself. It will trigger your sense of rightness and wrongness from the start. It's important to remember a few things walking into it:

First, there is not a universal agreed-upon sense of what is right and wrong across societies and cultures around the world. (This is a challenging thought, but we'll explore this more in a moment.) Second, each of us has a unique morality, shaped in part by our genes, our culture, and our unique set of experiences.

Moral Foundations Theory is a model developed by Haidt and colleagues to explain these origins and variations in human moral reasoning. It is an observational device—a kind of metaphorical tool—through which to look at the diverse human experience of determining what is right or wrong.[9]

Moral emotions are profoundly deep signals for us as social animals. Since they define the boundaries of our group's norms, we cannot update them easily. If they were easily altered, it might cause us to be ostracized from our group, which was often fatal in ancient times. This is why unpacking moral emotions is so hard. We don't often update our moral intuitions in the face of evidence—we'll usually just find a reason to not believe the evidence.

Haidt uses the metaphor of *taste buds* to explain people's differing moral foundations. I might have a slightly different preference for sweet things or salty things. You may have an aversion to spicy foods, while I like them. Moral emotions operate similarly. You may be born with a more extreme aversion to betrayal from a friend, or an inherent sense of disgust when someone doesn't show respect to an elder. I may have a natural disposition to be enraged by seeing someone treated unfairly, while your colleague may not be upset by it quite as much. This is your moral palate.

Haidt and his colleague Jesse Graham created a questionnaire to evaluate people's different moral foundations. You can take it online at https://yourmorals.org to see what your personal moral foundations are.

They can be broken into six separate groups. Care, Fairness, Authority, Ingroup, Purity, and Liberty. Conservatives and liberals tend to have dramatically different moral foundations.

To illustrate this for friends around the dinner table, I will often ask a group of people to close their eyes and answer a series of questions adapted from *The Righteous Mind*. If they feel strongly that something is morally wrong, I tell them to raise their hands.

"A woman is cleaning her bathroom, and has run out of rags. She finds a flag—the flag of your country—and decides to use that to clean the various parts of her toilet that need to be cleaned. Does this feel wrong?"

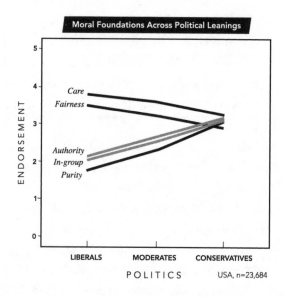

Pause. Some people raise their hands. Others don't.

"A brother and a sister go camping together—roughly similar in age and young, single adults. They are very close. During this camping trip, one night they decide to have sex. It is fully consensual, and they use two different forms of birth control. The next day, they wake up and decide this was a special experience they shared, but that they won't tell anyone else about it, and that it will not happen again. They go about their lives keeping this as their little secret. Does this feel wrong?"

For many, these questions feel *obvious* in their wrong-ness. What is surprising is how different people's reactions are, even at dinner tables among family members, lovers, or close friends. Many people don't think they have moral differences from their partners, and are surprised when they find out they do.

The reason I ask people to close their eyes is that there is a *powerful social force* in a room dedicated to calibrating moral issues through group referencing. If eyes are open, most people quickly look to their compatriots and try to figure out what's "right" in the space around them, superseding their moral instincts. It's as if their moral emotions are looking to sync up with others as they cocreate the moral matrix around the dinner table. This is the same social force that is present when we are presented with moral issues on social media, looking to our friends and followers to determine what is right or wrong.

What's more, when I ask my friends why they answered a certain way—why it was wrong—they do their best, but often struggle. I ask them to notice the strange tug and impulse to "attach a reason" to this powerful underlying feeling they might have about the rightness or wrongness of a particular answer, even if there's no fundamental logic beneath it. Usually, they find themselves ending up with an answer that can be summed up as "because it doesn't feel right." Haidt refers to this as moral dumbfounding.

The *why* doesn't need to exist for the emotion to be present. The emotion exists independent of the reason.

Let's return to the model of System 1 and System 2 processing we discussed earlier—fast and slow thinking. We quickly make a moral judgment using the former, and rationalize it after the fact with the latter. Haidt describes these two systems forces at play with another useful metaphor: an elephant (reactive System 1), and a rider (rational System 2). The elephant is the emotional impulse that we have, a large and powerful beast moving in whatever direction its feelings pull it. The rider is our logical brains, our System 2 processes, attempting to *justify* the movements of our emotional System 1 elephants. The rider is ostensibly a clever lawyer sitting on the back of this emotional beast, exclaiming to the world exactly *why* we believe what we believe. But the elephant is almost always in control, providing the emotional inertia pulling us toward a judgment, which our very logical brains then explain away.[10]

This is a helpful exercise for people who regularly find themselves in difficult political disagreements. When we disagree about moral issues, usually we're not disagreeing constructively. We're almost always explaining the *why*, letting our riders do the arguing, not addressing the feeling. But the problem with this is that when our riders are doing the talking, they are not convincing the elephants. They are not attempting to appeal to the moral foundations of the listener. This is one of the reasons why it's so difficult to change people's minds when it comes to political issues. Our careful lawyers are sitting there upon our elephants, with anecdotes and figures and takedowns they've heard before, ready to do verbal battle.

But if you want to actually change someone's mind rather than just score points, pay attention to the elephant. Address the emotional foundations beneath the lawyer. Do your best to ask questions about how the transgression feels. Appeal to the elephant's needs. You'll make far more progress.

Understanding the moral foundations of the person with whom you are debating you can actually begin to speak to their elephant. Thinking through our moral foundations can be a helpful frame if you want to persuade, rather than condemn the people you disagree with—a key element missing from the way we communicate with others online.

These moral emotions help advance group dynamics. As Haidt says, "Moral emotions bind and blind."[11] They encourage a sense of group identity. They *bind* us to the beliefs and sacred values of the group, and will often *blind* us to any information or evidence that challenges the group's worldview.

Collectively, these morals adhere to create cultural norms that we follow within our communities. This collective moral knowledge exists inside what Haidt calls a *moral matrix*—a network of shared norms, meanings, principles, and concerns that "provides a complete, unified, and emotionally compelling worldview."[12]

This time period we were born into has a number of competing moral matrices, largely based on the moral foundations of our communities. Understanding our moral matrix is a huge part of understanding why and how social media is changing the way we see the world.

Where Moral Emotions Come From

If we have moral foundations, does that not suggest they're unchangeable? And where do they come from? Are they embedded within us from birth?

The word *foundations* itself implies immutability, a stone base that the structure of our worldview is built upon. But this language is imprecise when we consider the organization of our brain from birth.

Our morality comes built-in, but it's not written into our genes. A baby isn't born with an immutable moral code. As the neuroscientist Gary Marcus says, " 'Built-in' does not mean unmalleable; it means 'organized in advance of experience.' "[13]

So, our morals aren't operating instructions carved into a stone tablet. Instead, when we're born, our morals are more like an open word-processing document with a set of clear instructions hard-coded at the top. An incomplete draft in a rough outline of what we might become. This document, as we grow, is updated and changed based on our lived experiences. Bits are deleted, bits are replaced. What we're left with is a moral code that guides us deeply and emotionally on a daily basis.

There are many instances of kids born into ultraconservative families who switch to liberal politics as they reach adulthood. And vice versa. You may have friends who were born into liberal families, only to find themselves drawn to more conservative positions as they grow older. Some research on this actually shows that the more emphasis placed on politics in a child's life, the more likely they are to establish different political preferences as they grow up.[14]

If this baseline of moral foundations sounds dismal for our ability to sense make in a multiethnic, multifaceted pluralistic society, I understand your worry. But this disposition toward moral tribalism is a bit of a problematic idea. It implies that we have *one* tribe that we belong to. In fact, we usually have many.

The Magical Band

Imagine carrying around a magic elastic band. Upon your command, it shrinks or expands in your hands. You can, whenever you want, instantly use it like a lasso and toss it around a group of humans and make them care about one another and treat each other more like kin. You can stretch this elastic band around any number of humans as long as they share a common set of labels, experiences, or a common enemy.

This is what "identity" is: a magical band we can use to feel close to others. Moral emotions are part of the rubbery material that composes this magical band.

Beyond just the signal-processing value of looking for social cues to determine and find additional information, moral emotions go one step deeper. They help us survive and thrive within the identity of a group and community.

In millennia past, we would have found ourselves in single, semicohesive groups of several dozen hunter-gatherers. In modern times, we find ourselves in many elastic groups with identities that overlap in and among many other groups of people. This intergroup flexibility is a huge advantage for a highly complex society like the one we live in now.

Many billions of humans interact daily, with interlaced identities and hierarchies. These allow us to play different games with our magical band of identity at different times. Two theories, social identity theory and self-categorization theory, describe how individuals stretch these lines of group membership into their identities.[15] Together they show how our identifications are quite malleable depending on the situation.

Your identity might be Catholic (1.3 billion humans), American (320 million humans), Republican (50 million humans), African American (42 million humans), and from Houston (10 million humans). You might also cheer on your alma mater's football team (60,000 humans), and be a member of your local Rotary Club (140 humans), and play on your local softball team (10 humans).

Each of these different groups is distinct, and you may be a part of several. But each can provide you with a sense of identity and camaraderie that's fundamentally helpful for your sense of belonging, personal advancement, and well-being. Each provides a sort of privileged network of connection—one that can be used to help parse information and deal with the complexities of life.

Threats Harden Our Identities

But while our identities are often malleable, when they are threatened they become much more rigid. Think of an explicit attack on one of these groups, such as the terrorist attacks on 9/11. Overnight, the collective identity of all Americans became far more visible. Many felt the need to display flags where they lived, and shows of patriotism celebrating American values became widespread. In addition, the identity of New Yorkers, the primary location of the attacks, became an even stronger and more salient identity. When under threat, the elastic band of these identities became rigid and inflexible.

When our in-groups are threatened, several moral emotions become powerful tools for maintaining the sanctity of these identities: outrage, contempt, and gratitude. We want to make sure our in-group is better than the one threatening us. We point our outrage and contempt at those out-groups who threaten our identities, and gratitude toward our in-group signaling its value.

Social media does especially strange things to this magical band. It allows us to see threats where we previously did not see them, which in turn causes our identities to thicken.

A 2016 study showed that the vast majority of Facebook and Twitter users (94 percent on Facebook and 89 percent on Twitter) reported that when they log on to social media they see at least some political content. And users are far more likely to see morally or politically charged content—content that evokes outrage—on social media compared to other sources. This outrageous content is often directly related to cultural, racial, gender, sexual, and political identities.[16]

When we see people attacking a group we identify with or care about ("Republicans did this disgusting thing!" or "Democrats are wrecking America!"), it forces us to affix our own position in relation to those groups. These identities become far more visible and prominent as we feel called to defend them or condemn them. Like a steady heat applied to soft plastic, the constant flow of opinions and threats from our news feeds cause the flexible bands of our identities to harden in place.

This doesn't just harden an abstract sense of personal identity. It actually changes how we determine what is true.

Each of our identities has shared values, ideas, and convictions that help us make sense of things. Every group you are a part of has a basket of shared beliefs, which you implicitly agreed to when you joined it. While it's often helpful to rely upon these beliefs when we're making sense of life's complexity, it's very unlikely that all of these beliefs are correct. We started this chapter learning that we make better sense of the world when we have multiple viewpoints. When we allow for our intuitions to be challenged, we can do much better at determining what is accurate. As we'll explore in part III, larger groups and organizations—institutions like journalism and academia—were created to formalize some type of viewpoint diversity, with systems in place to challenge our individual and tribal assumptions. Without such diversity, and largely as a result of social media's influence on our discourse, we've ended up defaulting to identity-based narratives that reduce our capacity to discern the truth. To paraphrase Haidt, "When a group lacks viewpoint diversity, it becomes structurally stupid."[17]

IN SUM

We began this chapter learning how a single video of a terrible incident with pit bulls polarized me on the breed. I was able to confirm my new opinion with a simple search on the internet. I only began to reconsider that belief after being challenged by a friend to examine my perspective, which allowed for me to update my views more accurately.

- The social intuitionist model explains how this happens, and shows how feelings of moral "rightness" or "wrongness" happen almost instantaneously, with our justifications coming later.
- Social interactions with others can help us challenge these intuitions. But the dynamics of social media often push us into a reinforcing feedback loop, where friends often reinforce our intuitive beliefs. Our intuitions about what is right and wrong come from moral emotions.
- Moral emotions are not universal, and because of this, a sense of right and wrong are not universally shared. According to Jonathan Haidt, everyone has a unique set of "moral taste buds" that derive from (at least) six foundations: care, fairness, loyalty, authority, liberty, and sanctity. Liberals and conservatives have different quantities of each of these.
- We have these moral foundations for a reason: They allow for us to quickly make sense of violations in our communities, and help us cohere into groups with shared identities.
- But our identities are not etched in stone. Instead, they operate like a "magical elastic band" that provides us with a privileged network of connections to help us share in the hardships and complexities of life.
- These identities become more rigid and more salient when they are threatened. Social media and frightening news often cause us to see

identity threats where we previously did not see them, *forcing our identities to harden in place.*

Let's now take into account all we've just learned about our moral emotions and about our disposition to form tribal identities, with all of their strengths and pitfalls. What if we could design a giant room to mess with them? What would it do? Let's find out.

Chapter 12

The Worst Room

What you have said in the dark will be heard in the daylight, and what you have whispered in the ear in the inner rooms will be proclaimed from the rooftops.

—*Luke 12:3*

Imagine walking into a room with four billion humans in it.

The room is enormous: fifty miles to a side, with a tall, vaulted ceiling, and walls the height of a twenty-story building. Inside, people are standing shoulder to shoulder, conversing with each other. The room is filled with a constant background din of billions of voices chatting.

As you walk inside, you're given a tiny microphone. It takes a moment to get your bearings in this vast and noisy space, but once you've aligned yourself you notice two strange things: First, everyone else has a microphone and is speaking into it. The next thing you notice is a massive marquee on the far wall. Above the marquee is an enormous sign that reads, "The Most Important Sentence." Everyone is facing it.

As you watch, the marquee lights up and a huge phrase is splashed across the wall, scrolling word by word. As you read it out, you feel a knot of anger in your throat and a flush of red in your cheeks. It's a righteous and outrageous sentence, something terribly insensitive, truly offensive.

The room shifts around you, and you hear millions of other people reacting the same way, muttering as disgust and anger ripple through the crowd. The tone of the room has changed. The background din is now louder, more heated.

A few moments pass and the marquee lights up again. What scrolls past now is a hyperbolic response to the previous sentence—also totally ridiculous—an extreme statement that's even more outrageous.

It says, "IF YOU DON'T AGREE, YOU ARE EVIL."

And in the next beat:

"IF YOU AGREE, SHAME ON YOU."

You feel another wave of disgust as the words scroll by.

Swiftly, you realize that everyone's microphones are on, and by some hidden program the most outrageous statements, the most disgusting ideas, and the foulest anecdotes—selected from the billions of humans in the room—are being splashed across the far wall for all to see. The most offensive comments are being captured and put on display, followed by the most extreme retorts.

You're angry, confused, and outraged. But you're also rapt, unable to look away. "This is ridiculous," you blurt out loud. Suddenly, the microphone in your hand buzzes. You look down and see that a small readout on the side of the device shows a set of numbers. It says "4 points!"

Looking around, you realize every human in the room has a score counter—a tiny set of numbers attached to each and every thing they say. The more attention the phrase captures, the higher this number ticks. With every comment, they compete to occupy the Most Important Sentence. The room is a giant game, and everyone inside is playing for points.

How could this be so bad? How can a tiny set of numbers keep this many people playing? Each little thing they say captures a little bit of attention. Surely there are a lot of *good* things being said in the room—why aren't you seeing more of those on the giant wall? Something strange is happening here.

When you look at the counter, the scores on their surface seem fine. The description on each number reads as follows:

Score 1: How many people appreciate what you just said?

Score 2: How many people value what you just said enough to tell other people about it?

Score 3: How many people are following what you say?

All these check out. Pretty straightforward. But why on earth is the most important sentence so full of toxicity and outrage? Is the game rigged? What is wrong with these metrics? The metrics themselves have done something very strange to the people in this room. Let's explore how they work.

The Great Measuring Game

What gets measured gets managed.

—Peter Drucker (misattributed)

What gets measured gets managed—even when it's pointless to measure and manage it, and even if it harms the purpose of the organization to do so.

—Simon Caulkin, The Rule Is Simple: Be Careful What You Measure

I love metrics. I count the number of pull-ups that I do just about every day, and I have a fitness tracker that tells me how well I sleep at night. Every day I track my meditations in an app that gives me a score. It has a "run streak" that I aspire every day to maintain because I feel better when I meditate. Metrics help me make sense of my overall behavior, and focus on what to improve.

Putting a number on something is an innocuous enough act. These metrics are a way of externalizing order, of counting the world. Our brains are wired to put constraints on things, to separate them into divisible parts.

Many forms of human activity depend upon the pastime of attaching numerical values to objects, behaviors, people, and phenomena. Metrics are a fundamental part of the modern world. Business, trade, economics, and science would not be possible without measurement. The concept of "progress" itself requires some way of measuring it. Put simply, we understand the world by quantifying it.

But for all of this positive value that measuring allows, the act of quantifying something can change it for the worse.

In 1902, in Hanoi, Vietnam, an enormous rat infestation caused the colonial French government to launch a bounty program to reward rats killed in the city. The bounty was given for each severed rat tail turned in.

The program seemed to be working well, and many local Vietnamese began delivering rat tails to claim their reward money. After a period of time, however, the rat problem did not seem to be improving. Confused, when the authorities began investigating, they found that many of the city's street rats had lost their tails. The rat catchers would capture rats, remove the tails for the bounty, and release them into the sewers where they could continue to reproduce. The measure and the bounty had created a job: the lucrative new profession of rat-tail harvester.[1]

In business, measuring the wrong thing can also make bad things happen. A company that measured the number of customer service calls completed per representative, for instance, found that it easily backfired. Employees would cut short the length of their conversations with customers to bank more calls, often to the consternation of customers whose problems weren't adequately solved. (If they called back again, even better for the employees' numbers!)

Metrics are a type of abstraction. They flatten complex things into simple things. Because of this, every time a behavior is turned into a number, pieces of that behavior are lost. For many things, this is fine and advantageous. But that's not always the case with social metrics. In the quest for validation on social media, we may be unwittingly becoming the online equivalent of rat-tail harvesters.

The Creation of Social Metrics

Over the course of two winter days in 1968, a group of strangers assembled on the corner of a city block in New York City. They were all doing something odd.

At seemingly random times, they would, in unison, look upward to the corner of a building above the street. Onlookers stared, collectively rapt, like something exciting was about to happen. Large numbers of passing pedestrians would see the group of gawkers and stop to look up, squinting overhead to see what the fuss was about, searching for the show worthy of their attention. But nothing happened.

This was an experiment. The group of strangers staring at the upper floors were actors. A researcher stood nearby, taking note of how many people walking past on the street stopped what they were doing to look upward with the group. Many did. A majority of passersby also turned their gaze skyward to figure out what this gaggle of common strangers was looking at.

These unwitting strangers were all part of one of the most consequential series of studies on human conformity. Stanley Milgram, the study's author, was curious about when we take social cues from others, and how it affects our perception of events. In another experiment by Solomon Asch, actors standing in an elevator between floors would turn—in unison—toward the wall, and count how often strangers entering the elevator would follow the group's silly behavior (many of them did).[2]

These experiments illustrated just how likely we are to follow the signals and behaviors of those around us. Our natural impulse toward seeking guidance from others is deeply ingrained in us as social creatures.

We know it intuitively: If a bunch of random people are looking at something, that thing is probably worth our attention. Social attention is a powerful signal. It's also one that has been amplified to an extreme degree online.

The Positives of Social Metrics

The act of clicking Like on someone's post feels quite good. It's an opportunity to provide a positive emotion to someone in an online space that requires minimal effort on your end—one with a sizable impact for them. Giving someone a like is an inexpensive act, but quite valuable for the recipient. On its surface, it's a win for everybody. Indeed, the creator of the Like button, Justin Rosenstein, said as much: "The main intention I had was to make positivity the path of least resistance."[3]

Having an open way of rating content is inherently democratic on its surface. The egalitarian nature of social media allows for anyone to say almost anything, and be judged by the perceived value of the content on its own. And that value becomes a signal that can be used to promote the content even further. No gatekeepers. No favoritism.

This is a sea change from the way that content propagated beforehand—presentation was largely determined by publishers, editors, curators, or censors, who would platform (or deplatform) ideas or a piece of content. Having social metrics such as a like or a share attached to content provides a way to evenly measure the playing field, and in doing so removes a piece of content's success or failure from the precocious whims of governing bodies.

But what is lost when such a simplified measure is applied to social discourse?

The Dangers of Social Metrics

I don't post to social media often, largely because I feel a strange force tugging at me whenever I do. I feel a sense of disappointment, anticipation, or satisfaction depending on the engagement my content receives. The number of likes and shares for a certain type of content makes me feel like I've found some microvalidation of my self-worth. A post that doesn't get much engagement intimately feels like a negative judgment of my value offered to the world. Most people who post to social media feel this momentary apprehension, like some strange algorithmic test of personal character.

There's a reason for this, and it goes back to a series of famous experiments carried out in the 1930s.

The psychologist B. F. Skinner was sure there was something unique happening in the minds of animals when they were being trained to do things. By placing pigeons in light and sound-dampened chambers, with a small lever available to them, he tested whether he could reinforce certain behavior (lever pressing) over time by giving the pigeons a stimulus (a flashing light) and then reinforcing it with a reward (food pellets) every time they pushed it.

The results were clear, across many species, from rats to monkeys, even to invertebrates—rewards teach us to repeat behaviors and create learned habits. He demonstrated that he could get animals to repeat a behavior if they received rewards for it. Skinner called this device an Operant Conditioning Chamber, which became known colloquially as simply a Skinner Box.[4]

There were three amazing discoveries in his research: The first was that Skinner Boxes work on humans just as well as other animals.

The second was that different types of rewards change how long we'll keep playing. If you give out predictable rewards with every action, eventually people (and animals) become less interested in them over time. If instead you add some pure randomness to the reward, they will keep playing far longer. These are called *intermittent variable rewards*, and they hack our reward pathways to keep us playing (see slot machines in casinos, loot crates in video games).

The third was that primary conditioners, rewards that involved basic biological needs (like food) had diminishing returns. Once that basic biological need is satisfied, people will stop playing. He instead found that so-called secondary reinforcers, like money, points, or social affirmation, don't hit a satiation threshold. People will just keep playing to try to rack up the numbers. In a very real way, B. F. Skinner discovered how to "hack" human behavior.[5]

Social Media Is a Skinner Box

Many reward types that capture and reinforce our behavior on social media work exactly like a Skinner Box. When a particular post gets a lot of engagement, it influences us. It gives us back validation for that specific content type. For example, if we get a lot of likes, shares, and comments from a post about kittens, we're much more likely to make that type of post again in the future. Fortunately, being flooded with photos of kittens isn't particularly problematic for society.

But being flooded by angry and incendiary posts is. In 2021, Yale researchers Molly Crockett and William J. Brady set out to figure out if Skinner's research applied to one of social media's more problematic effects: why people seem to share so much outrageous stuff online.

They put together a team that built a tool capable of tracking and measuring moral outrage on Twitter posts. They tracked 7,331 Twitter users, and combed through 12.7 million tweets, trying to determine how and why users were expressing moral outrage on the platform.

What they found was alarming. The users who received more likes and retweets when they expressed outrage in their post were *significantly* more likely to express outrage again later on. Looking for more evidence, the researchers followed this up with a series of behavioral experiments which showed that when users were explicitly rewarded for expressing outrage with metrics, it increased their *total number* of outrage expressions in the future. The rewards of social media clearly shift people's behavior toward moral expressions of anger and contempt.[6]

Even more surprisingly, they discovered that the groups more susceptible to these feedback incentives weren't radical or extreme political partisans, they were actually moderates. Crockett explained further:

> Our studies find that people with politically moderate friends and followers are more sensitive to social feedback that reinforces their outrage expressions. This suggests a mechanism for how moderate groups can become politically radicalized over time—the rewards of social media create positive feedback loops that exacerbate outrage.[7]

This research shows a clear pathway for moderate social media users to be rewarded for more displays of outrage. This is the most worrying implication of this research: If they had measured an earlier set of users, they might have found that today's radical users were yesterday's moderates. In other words, the longer the users spent on social media, the more likely they were to become politically extreme.

Followers

On its surface, the number looks benign: a follower count. A simple way of judging people on their value to others.

Building a metric for our popularity would seem to be an easy proxy for understanding who is and is not important—a time-saver for those who are interested in figuring out who is most likely to share the best information.

But participation is no longer really a choice. When we start an account on social media, we're instantly embedded in a comparison game: Who has more followers? Where do I stand? Who counts? Metrics like this push us into inescapable contests that we cannot ignore.

We automatically apply ourselves to the social ranking implicit in the system. By design this ranking is difficult to resist. These numbers overlay a type of popularity contest that reinforces use of the platforms themselves—and infiltrate our offline evaluations of others. The perks of this influence are wonderful for those who have many followers. Esteem comes quickly to those who pay attention to these numbers in real life. ("Did you hear, they have twenty thousand followers on Instagram?") Kids are inherently aware of the rewards available to them. One recent study showed that 54 percent of American young people between thirteen and thirty-eight would become a social media influencer if given the opportunity.[8]

We humans are highly calibrated sensors of our social environments. The psychologist Mark Leary coined the term *sociometer* to describe the inner mental gauge that tells us, on a moment-to-moment basis, how we are perceived by others. We don't really need self-esteem, Leary argued; rather, the evolutionary imperative is to get others to see us as desirable partners. We need social esteem. Social media, with its displays of friends and followers, has pulled our sociometers out of our private thoughts and posted them for all to see.[9]

The Hierarchy of Disagreement

In a 2008 essay, just as Web 2.0 was getting going, the programmer and entrepreneur Paul Graham noted that a natural by-product of spending more time online would be more disagreement. Being exposed to more opinions of others would simply give one more opportunities to argue. He offered a model he called the Hierarchy of Disagreement. He classified argument types into a seven-level scale.

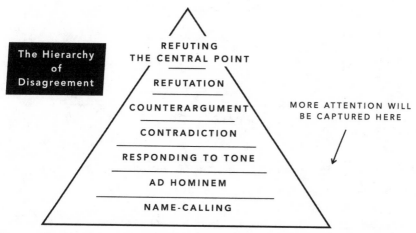

The Hierarchy of Disagreement

REFUTING THE CENTRAL POINT

REFUTATION

COUNTERARGUMENT

CONTRADICTION

RESPONDING TO TONE

AD HOMINEM

NAME-CALLING

MORE ATTENTION WILL BE CAPTURED HERE

Today there are more incentives to push conversations down the hierarchy.

He noted that "moving up the disagreement hierarchy will make most people happier by reducing their meanness." At the very top, we have the refutation of the central point of an argument.

If we look at this hierarchy, we can see where failure modes might exist, particularly when there are metrics attached to our discourse. A vitriolic disagreement—one that appeals to your audience's inherent biases—might actually get you more followers than a productive disagreement. In an attention economy, recognizing the nuance of the other person's opinion is less likely to make a splash. Moving up the pyramid does little for your follower count.

Forced further down the hierarchy, arguments are stunted. In this new discourse, everybody loses access to the most valuable disagreements.

We might map this out on a graph, with one axis being Desire to Find the Truth, the other, Number of People Watching. As our audience increases in size, the desire to "win" an argument supersedes the desire to find the truth.

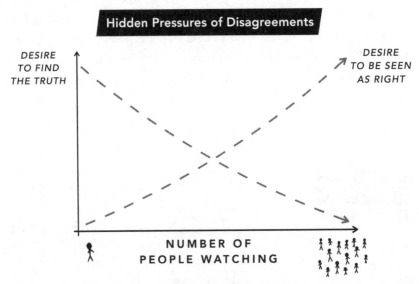

As the number of people watching increases, the desire to find the truth decreases in proportion to the desire to be seen as right.

The researchers Justin Tosi and Brandon Warmke coined the term *moral grandstanding* to describe when people use moral language to boost their reputation.[10] With a debate happening in a public forum, every orator tries to outdo the previous one, resulting in a show of moral attitudes. In this competition to gain the approval of the audience, grandstanders often make up moral charges, pile on in cases of public shaming, and state that anyone opposing them is obviously wrong. They will often exaggerate their emotional displays to appeal to their audience. These moral grandstanders examine every word spoken by their opponents for the potential to evoke public outrage. The speaker's intent is ignored.[11]

Because of the sheer number of observers and our tendency to seek signals from our online communities, many of our disagreements on social media become metrics-driven opportunities for grandstanding. These disagreements don't often trend toward productive exchange. Instead, they become opportunities for polemics and showmanship, tactical battlegrounds rather than opportunities for learning. When others are ranking and scoring us in real time, we lose the ability to examine new concepts in good faith.

This Game Is Not Just a Game

Social media can seem like a game. When we open our apps and craft a post, the way we look to score points in the form of likes and followers distinctly resembles a strange new playful competition. But while it *feels* like a game, it is unlike any other game we might play in our spare time.

The academic C. Thi Nguyen has explained how games are different: "Actions in games are screened off, in important ways, from ordinary life. When we are playing basketball, and you block my pass, I do not take this to be a sign of your long-term hostility towards me. When we are playing at having an insult contest, we don't take each other's speech to be indicative of our actual attitudes or beliefs about the world."[12] Games happen in what the Dutch historian Johan Huizinga famously called "the magic circle"— where the players take on alternate roles, and our actions take on alternate meanings.[13]

With social media we never exit the game. Our phones are always with us. We don't extricate ourselves from the mechanics. And since the goal of the game designers of social media is to keep us there as long as possible, it's an active competition with real life. With a constant type of habituated attention being pulled into the metrics, we never leave these digital spaces. In doing so, social media has colonized *our* world with its game mechanics.

Metrics Are Money

While we are paid in the small rushes of dopamine that come from accumulating abstract numbers, metrics also translate into hard cash. Acquiring these metrics don't just provide us with hits of emotional validation. They are transferable into economic value that is quantifiable and very real.

It's no secret that the ability to consistently capture attention is an asset that brands will pay for. A follower is a tangible, monetizable asset worth money. If you're trying to purchase followers, Twitter will charge you between $2 and $4 to acquire a new one using their promoted accounts feature.

If you have a significant enough following, brands will pay you to post sponsored items on their behalf. Depending on the size of your following in Instagram, for instance, these payouts can range from $75 per post (to an account with two thousand followers), up to hundreds of thousands of dollars per post (for accounts with hundreds of thousands of followers).[14]

Between 2017 and 2021, the average cost for reaching a thousand Twitter users (the metric advertisers use is CPM, or *cost per mille*) was between $5 and $7. It costs that much to get a thousand eyeballs on your post. Any strategies that increase how much your content is shared also have a financial value.[15]

Let's now bring this economic incentive back to Billy Brady's accounting of the *engagement value* of moral outrage. He found that adding a single moral or emotional word to a post on Twitter increased the viral spread of that content by 17 percent per word.[16] All of our posts to social media exist in a marketplace for attention—they vie for the top of our followers' feeds. Our posts are always competing against other people's posts. If outraged posts have an advantage in this competition, they are literally worth more money.

For a brand or an individual, if you want to increase the value of a post, then including moral outrage, or linking to a larger movement that signals its moral conviction, might increase the reach of that content by at least that much. Moreover, it might actually improve the perception and brand affinity by appealing to the moral foundations of the brand's consumers and employees, increasing sales and burnishing their reputation. This can be an inherently polarizing strategy, as a company that picks a cause to support, whose audience is morally diverse, might then alienate a sizable percentage of their customer base who disagree with that cause. But these economics can also make sense—if a company knows enough about its consumers' and employees' moral affiliations—it can make sure to pick a cause-sector that's in line with its customers.

Since moral content is a reliable tool for capturing attention, it can also be used for psychographic profiling for future marketing opportunities. Many major brands do this with tremendous success—creating viral campaigns that utilize moral righteousness and outrage to gain traction and attention among core consumers who have a similar moral disposition. These campaigns also often get a secondary boost due to the proliferation of pile-ons and think pieces discussing these ad spots. Brands that moralize their products often succeed in the attention marketplace.

This basic economic incentive can help to explain how and why so many brands have begun to link themselves with online cause-related issues. While it may make strong moral sense to those decision-makers, it can make clear economic sense to the company as a whole as well. Social media provides measurable financial incentives for companies to include moral language in their quest to burnish their brands and perceptions.

But as nefarious as this sounds, moralization of content is not always the result of callous manipulation and greed. Social metrics do something else that influences our behavior in pernicious ways.

Audience Capture

In the latter days of 2016, I wrote an article about how social media was diminishing our capacity for empathy.[17] In the wake of that year's presidential election, the article went hugely viral, and was shared with several million people. At the time I was working on other projects full time. When the article took off, I shifted my focus away from the consulting work I had been doing for years, and began focusing instead on writing full time. One of the by-products of that tremendous signal from this new audience is the book you're reading right now.

A sizable new audience of strangers had given me a clear message: This was important. Do more of it. When many people we care about tell us what we should be doing, we listen.

This is the result of "audience capture": how we influence, and are *influenced by* those who observe us. We don't just capture an audience—we are also captured by their feedback. This is often a wonderful thing, provoking us to produce more useful and interesting works. As creators, the signal from our audience is a huge part of why we do what we do.

But it also has a dark side. The writer Gurwinder Boghal has explained the phenomena of audience capture for influencers illustrating the story of a young YouTuber named Nicholas Perry. In 2016, Perry began a YouTube channel as a skinny vegan violinist. After a year of getting little traction online, he abandoned veganism, citing health concerns, and shifted to uploading *mukbang* (eating show) videos of him trying different foods for his followers. These followers began demanding more and more extreme feats of food consumption. Before long, in an attempt to appease his increasingly demanding audience, he was posting videos of himself eating whole fast-food menus in a single sitting.[18]

He found a large audience with this new format. In terms of metrics, this new format was overwhelmingly successful. After several years of following his audience's continued requests, he amassed millions of followers, and over a billion total views. But in the process, his online identity and physical character changed dramatically as well. Nicholas Perry became the personality Nikocado—an obese parody of himself, ballooning to more than four hundred pounds, voraciously consuming anything his audience asked him to eat. Following his audience's desires caused him to pursue increasingly extreme feats at the expense of his mental and physical health.

Nicholas Perry, left, and Nikocado, right, after several years of building a following on YouTube. Source: Nikocado Avocado YouTube Channel.

Boghal summarizes this cross-directional influence.

When influencers are analyzing audience feedback, they often find that their more outlandish behavior receives the most attention and approval, which leads them to recalibrate their personalities according to far more extreme social cues than those they'd receive in real life. In doing this they exaggerate the more idiosyncratic facets of their personalities, becoming crude caricatures of themselves.[19]

This need not only apply to influencers. We are signal-processing machines. We respond to the types of positive signals we receive from those who observe us. Our audiences online reflect *back* to us what their opinion of our behavior is, and we adapt to fit it. The metrics (likes, followers, shares, and comments) available to us now on social media allow for us to measure that feedback far more precisely than we previously could, leading to us internalizing what is "good" behavior.

As we find ourselves more and more inside of these online spaces, this influence becomes more pronounced. As Boghal notes, "We are all gaining online audiences."[20] Anytime we post to our followers, we are entering into a process of exchange with our viewers—one that is beholden to the same extreme engagement problems found everywhere else on social media.

IN SUM

We began this chapter walking into a giant room full of people linked up to a system that was measuring what they were saying. The room gave three key visible metrics back to them: followers, likes, and shares. The room strangely devolved into a competition to say the most salacious thing to score the most points. This is what we learned about the room:

- Visible metrics have the benefit of helping to even the playing field by removing gatekeepers in content promotion, but they also subtly influence our behavior. The visibility of these metrics resembles a digital version of what psychologist Mark Leary described as a "sociometer" illustrating the inner mental gauge that shows us, moment by moment, how others perceive us.

- When our audience increases in size, the desire to "win" an argument can override our desire to find the truth, pushing us down what Paul Graham refers to as the Hierarchy of Disagreement, toward ad-hominem attacks and away from constructive refutation.

- Vying for metrics on social media works exactly like an Operant Conditioning Chamber, otherwise known as a Skinner Box. When a particular post gets a lot of engagement, it trains us to post more.

- As we build our personal brands and platforms, we don't just capture an audience—we are also captured by their feedback. The dark side of this audience capture emerges when we lose our authentic identity in the pursuit of their approval.

- Oftentimes, these audiences respond well to moral and emotional content. Since this approval can be monetized, individuals and brands can financially benefit from taking strong stands on divisive cultural issues.

Next up, we'll explore what happens when the metrics-driven pursuit of audience approval crosses a line, and the audience turns against the creator.

Chapter 13

Trauma, Processing, and Cancellation

In March of 2014, a twenty-three-year-old writer named Suey Park opened Twitter to see a post on her timeline from the account of *The Colbert Report*. The tweet was an attempt at a joke, making fun of the newly launched Washington Redskins Original Americans Foundation, which had been recently announced by Daniel Snyder, the team's owner. The joke had been made by Stephen Colbert on his evening show, and posted by a Comedy Central employee to the show's Twitter account, and said, "I am willing to show #Asian community I care by introducing the Ching-Chong Ding-Dong Foundation for Sensitivity to Orientals or Whatever." The gag on the show was meant to be fairly straightforward, pillorying the absurdity of Snyder naming a foundation to support a marginalized group—all while using a derogatory term for the group in the name itself.[1]

The joke had been shared several times over the previous days on the show's other social media accounts, but always with a link to the original video from the show. This tweet, context free, did not sit well with Park, a Korean American.

Upset, she wrote on Twitter: "The Ching-Chong Ding-Dong Foundation for Sensitivity to Orientals has decided to call for #CancelColbert. Trend it."

Park went on to post a series of critical tweets on the platform directed at Colbert and encouraged others to do the same. Within a few hours #CancelColbert had, as Park hoped, begun to trend. But not for the reason she had hoped. Many thousands of Colbert fans, believing that Park didn't understand the original context of the tweet, came to his defense with their own outraged responses, causing the hashtag to trend upward. Park pushed back, insisting that the initial insensitive tweet was fundamentally wrong.

More people piled on, insisting that she was missing the point. This outrage *about* the outrage gave it the biggest boost, as dozens of mainstream news organizations, seeing the traffic spike, picked up the story. *USA Today*, *Slate*, *Variety*, the *New Yorker*, CNN, and the *Wall Street Journal* all ran features, along with three stories in *Time* and seven in *Salon*. By the end of the day, #CancelColbert became a top-trending item on Twitter. Very little of the outrage was directed at Daniel Snyder's unfortunate naming decision.[2]

This particular outrage cascade was the result of an attempt at satire, interpreted and magnified beyond proportion. It was also one of the earliest explicit uses of the word *cancellation* on social media.

Colbert's show was not canceled by the cascade of judgment against him. He would soon go on to be asked to take over hosting *The Late Show*, where he would shed his faux-conservative alter ego and embrace liberal causes far more openly.

Park started and facilitated a number of activist hashtags like #NotYourAsianSidekick, #POC4CulturalEnrichment and #BlackPowerYellowPeril. She amassed an account with a bit over twenty thousand followers in the process, but also received a large amount of negative attention for being an early hashtag activist. Eventually, Park herself took a step back from Twitter after the backlash to her full-contact social media activism took its toll.[3]

But in 2014 Park was perhaps just slightly too early for her flavor of activism to be taken seriously. #CancelColbert did not end Colbert's show, but it did help introduce a strange new phenomenon that would soon become a cultural force impossible to ignore: Cancellation. These events are now a fundamental part of our media ecosystem, and a core way we generate news—particularly partisan news.

This early cancellation is a strong example of how the social media news ecosystem thrives on confusion and outrage, showing how everyone gets paid along the way.

Park, a sensitive twenty-three-year-old looking to bring attention to a cause she cares about, interprets an offensive joke presented out of context as an injustice, and posts her interpretation. She is paid in likes, shares, and followers—and by bringing attention to the cause she cares about: racism against Asians.

Twitter users post comments to correct her what they view as outraged misinterpretation, by tweeting about the injustice's injustice. They are paid in likes, followers, shares, along with a shared camaraderie of defending a member of their in-group.

Several dozen publications see an opportunity to write about the confusion and outrage and amplify the fringe controversy *beyond* social media into the mainstream media. They cover the story because they know it will drive traffic, and they think it's an important moment. They are paid in actual dollars by advertisers who place ads on their sites.

The result was that everyone was paid in the process, but the world was worse off. Thousands of hours of misdirected attention and confusion, sitting on the back of an out of context piece of satire presented in social media. As Colbert sardonically quipped in the first episode back after the controversy, laying out what had just happened: "The system worked."

This particular moral moment illustrates a key relationship between trauma (a highly charged term today) and comedy. It's arguable that this was an important moment of cultural reflection on the role that Asians play in being the butt of other people's jokes. It's arguable that Park was wrong and was overreacting. It's also arguable that Colbert's writers crossed a line in making the joke in the first place. But this particular episode details what happens at the strange intersection of comedy and social media. It holds a few keys to understanding exactly how moral sensitivities can become system-wide explosions of emotional disgust.

As we unpack what happened here, I want you to note the emotions that come up for you when reading this examination. Reading about Colbert's joke, or Park's backlash, or the backlash to the backlash, you'll likely want to find an enemy in the narrative. Is it Park? Is it Colbert? Is it the news organizations that capitalized on the outrage? Is it the engineers who wrote the algorithm at Twitter? Is it society writ large for having little sympathy for the grievances of Asian Americans? Who is to blame?

Moral events like this are seen and felt by those who observe them. Just by examining this issue, you may notice a stirring of righteousness in your throat, or discomfort and anxiety somewhere in your body. That feeling is your moral matrix being updated. Just knowing someone somewhere is upset by an issue is enough to tighten your personal views on what is OK and not OK. This is a new type of moral tension you are now carrying with you.

This particular moral moment illustrates a key relationship between trauma, comedy, and cancellation in our online lives.

Tragedy Plus Time

The way we process bad things can be explained by an old adage about comedy that asserts "Comedy is tragedy plus time." This saying implies that there is a period of sensitivity after a tragedy in which things are not acceptable to joke about. Comedians love to skirt this line of acceptable bad things, because there's tension that can then be released through laughter. The comedian Hannah Gadsby has referenced this formula, noting that "punchlines need trauma."[4] Comics are known to play up the awkwardness of these uncomfortable moments with their audiences, eventually releasing it like a pressure valve, giving people permission to laugh.

In this way, the leftover tension from shared traumas can make effective comedy. If we were to map it out, it would look like this:

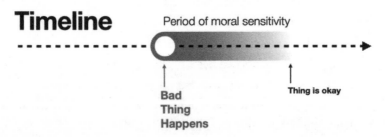

Illustration of "Comedy is tragedy plus time." As time passes, a visible traumatic event is followed by a period of increased sensitivity that eventually dissipates, after which it may be laughed about.

This was what Colbert was trying to do with his joke: turn a tragedy into a comedic moment.

Colbert's joke, using an outdated reference to Asians as "Orientals," was a pointed satire of a kind of explicit racism that used to be far more common in the United States and around the world. This racism was historically widespread but is much less explicit today. The joke subtext was pointed derision at the equivalent thing happening to Native Americans at that very moment. Presumably, Colbert's writing team in constructing the joke figured that this type of comedy was beyond the point of moral sensitivity. Joking about it in a time when people regularly called Asians such things would not be a joke—it would be crass and humorless—poor taste if it was construed as racism and not satire. At that point, in 2014, Colbert's joke (with the setup, and in context) was, for the majority of his audience, beyond the period of moral sensitivity.

But it wasn't that far off from the original trauma that some Asian Americans might experience, particularly those living in minority communities in certain parts of the country.

As an Asian American activist, Park was just too close to the joke to find it funny. She might have experienced extensive racial profiling in her life, she might have been nursing her own trauma. Regardless of the reason, she was angry about it, wanted to make a point, and had a powerful new platform through which to do it.

The first dimension of sensitivity is time, the other dimension that matters is social distance. If one person is exposed to a terrible thing, the people who care about them usually know not to make light of it because it might trigger them by bringing up old trauma again.

Oftentimes, sensitive individuals (people affected by the bad thing), in this case Park, feeling sensitive about the historical tragedy of overt and rampant Asian racism, can make a claim that connects to an old tragedy, even if they're unrelated (such as Colbert's joke posted out of context).

People who make these linkages are literally *calling out* the importance of the old trauma, and reminding others that they are still feeling sensitive, and that these feelings matter. This callout comes in the form of moral outrage, evoking the emotions of shame, disgust, or anger. Colbert's joke, and its subsequent backlash, trigger a new moment of cultural sensitivity, one that now hangs in the air. It's a sensitivity that now he (and others who observed the media explosion) will likely avoid in the future.

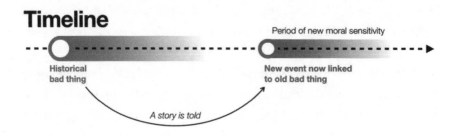

For activists and advocates, this process of revisiting old tragedies can be a way to make their ideas look especially important to pay attention to. It's a way of validating the significance of their experience in the eyes of others.

Journalists regularly do this, too, in the form of opinion pieces and analysis, attempting to bring context to an event to make it more interesting and weighty. This storytelling process is core to sensemaking. It's how we determine what might be a singular event or a pattern of events.

But with social media, any individual with a noteworthy personal grievance can do the same thing. Park was exhibiting behavior we learned about back in chapter 12: she was "grandstanding" in her condemnation of Colbert, and on the other side, many of her detractors were probably doing the same. It didn't matter if there was an indirect connection to the older trauma of systemic racism against Asians, because to her the connection *felt* real. Colbert was a powerful white man with a platform, probing an existing wound at her expense. She felt hurt, and it was important to her that other people knew it was hurtful.

When people feel emotions, regardless of the source, the emotions require acknowledgment. They don't need to be coddled, or encouraged, or amplified, but they do need to have the space to transit through the mind of the individual. What matters is that the feeling has a chance to be acknowledged so it can move along.

Instead, her emotional processing was encapsulated inside a new container—a tweet—where it was given an immense second life. This emotion, now living on its own, went on to ripple across the internet and trigger the emotions of many others.

Trauma and Processing

We can explore trauma and processing with a hypothetical story from ancient times. Imagine it's long ago, and you are walking down a path with your tribe. Mostly, the path is calm, and the walk is peaceful. One day, the path goes along a steep and dangerous ravine. As you traverse it, someone trips, loses their balance, and falls down the slope, seriously hurting themselves. They are badly injured. People carefully rush down to help them up, but it's too late to do much. The person is visibly harmed. A cloud of dust now sits in the air around the wounded person, making it hard to see. A lingering trace of a tragic moment, visible to everyone. No one laughs. The person is suffering, as are the people around them who witnessed the painful event. People stay for a while, helping them. But eventually the tribe must move on. The wounded person is now carried along, and people walk in silence, feeling their concern. Your tribe carries the burden together.

Much later, years after the tragic fall, the group is traveling along a similar path next to a gorge. Unexpectedly, you trip—but stumble and catch yourself. Someone, with perfect timing, makes a joke about the injured person and their unfortunate plight. You are the butt of the joke. It's slightly crass—the edge of acceptable. But it's been long enough that it's now okay. The injured parties have long healed. You can't help yourself—you laugh. Others laugh, too. Tears stream down your faces as you all laugh together. The tribe has moved on, and the trauma has been overcome. The air is clear. Things are okay again.

When a tragic event happens around us, it's an injury we all share— a moment in which we are collectively stunned and saddened. Events like a surprising accident, an unexpected flash of violence, a clear injustice, or a major natural disaster are painful for all present. Afterward, something is in the air—the experience lingers. In the wake of these events, we cannot joke about them, and comedy is often met with hostility. There is a certain duration of time society needs to process it to move on, to find a space to reflect. The event lingers until it's okay to look at with levity, retrospection, and humor. Hard things happen, and we share these things to cope. Eventually, we might laugh, but it takes time.

All tragedies fade. It's one of our greatest strengths being able to metabolize and move on from terrible things together. It's what we do best. Let's return to the previous diagram.

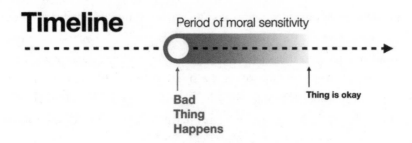

Timeline   Period of moral sensitivity

Bad
Thing
Happens

Thing is okay

When a bad thing is observed or experienced, it stirs up dust and emotional debris that hang in the air around us. Certain individuals and groups may be closer in proximity to the bad thing—or may have been literally wounded by it—which makes them understandably more sensitive.

Trauma is the Greek word for a physical, not an emotional, wound. In the mid-twentieth century it was extended to severe psychological wounds, the sort that soldiers in Vietnam experienced from observing or participating in horrific violence. In recent times it's become a catch-all word for any difficult emotional experience. The American Psychological Association defines it as "an emotional response to a terrible event," which has left it open to wide interpretation in culture and society.[5]

In modern times, we associate many things with trauma: traumatic relationships, traumatic childhoods, traumatic jobs, even traumatic memes.

But there are degrees of trauma. A violent rape is not the same as a betrayal by a friend. Being called a bad word is not the same thing as being physically assaulted by someone. Our legal system was built to help codify these degrees of harm, and prosecutors do their best to define it proportionally to the punishment for causing such offenses in others.

Traumas can be experienced by whole groups of people and societies. Shared traumas like war and disasters can deeply damage the fabric of a culture (though at times, shared trauma can actually strengthen people's ability to deal with them, such as during natural disasters or terrorist attacks). In Cambodia, for instance, just about everyone I met above a certain age showed some symptoms of what could likely be classified as PTSD. This was a natural reaction after surviving years of civil conflict and extreme poverty.

Trauma can be passed on to others in many ways. It can be passed from parent to children, in the form of abuse or learned anxieties. It can, to some degree, just be *felt* when we observe other people going through a traumatic event. It can also be passed on in relationships, when one person hurts another emotionally.

In relationships, for instance. Holly Muir and Spencer Greenberg have written about the clashing nature of traumas. Many serious relationships between friends or partners end because of these reactive events. One person does something benign that triggers a trauma of their partner, who then responds by taking a self-protective action, that in turn triggers self-protective counterreaction.[6]

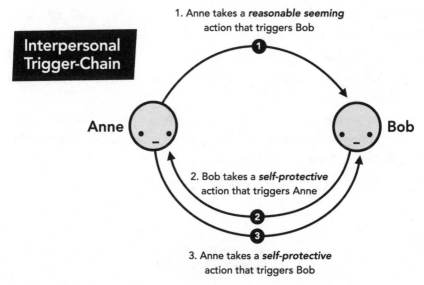

In a relationship, clashing traumas resemble a trigger-chain, in which one person's response triggers the others' trauma response. Adapted from Spencer Greenberg's graphic on Clashing Traumas, 2021

For example, Anne and Bob are in a relationship. Anne comments on something that Bob is wearing, saying it makes him look silly. Bob, having been teased as a kid, responds by going quiet and emotionally pulling away from Anne. Anne, who had an ex-boyfriend whose emotional distance presaged their breakup, protects herself by being terse with Bob. Bob, who now thinks that Anne is doubling down on her cruel behavior, pulls away even more. Both Anne and Bob have triggered each other's traumas in sequence, resulting in a chasm of emotional distance between them that both misattribute and neither understands.

If this sounds familiar, it's because it's another form of a trigger-chain, just between two people. One person's trigger sets off a trigger in the other person, who responds with a further triggering action. These interpersonal trigger-chains can be dangerous and damaging if they aren't identified and addressed in the moment.

Dealing with Difficult and Traumatic Emotions

A key feature of trauma is that it cannot be "fought." You will not win an argument with someone's trauma. Even if you feel like another person's trauma is overblown, incorrect, or out of bounds, you cannot meet it with anger, guilt, or shame and expect it to be reasoned with. This applies to our own difficult emotions as well.

As the psychologist Tara Brach has written, intense emotions are a sort of alert, telling us something is deserving of our attention. Sometimes letting an emotion be somatically felt, and removing all pretext and story from the expression of the emotion, can simply let the feeling move along.[7]

Let's return to Haidt's intuitionist model described in chapter 11.

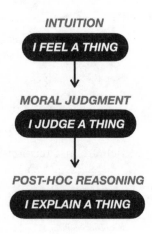

INTUITION

I FEEL A THING

MORAL JUDGMENT

I JUDGE A THING

POST-HOC REASONING

I EXPLAIN A THING

Oftentimes, when we try to process the difficult emotions we're feeling, we end up judging them as good or bad, or explaining them away. As the intuitionist model shows, both judgment and reasoning are separate events that happen *after* we feel an emotion or intuition. Both of these secondary processes aren't helpful for understanding *why* these emotions occur in the first place.

But there are several strategies for understanding difficult emotions. Both Gestalt therapy and Cognitive Behavioral Therapy put tremendous emphasis on *just feeling* an emotion, without attaching judgments. This has been put into practice by mindfulness practitioners like Michelle McDonald, who have codified a set of steps one can take to understand emotions as they move through us, which she calls RAIN.[8]

Recognize what is happening;
Allow the experience to be there, just as it is;
Investigate with interest and care;
Nurture with self-compassion.

This can be hard work, especially around particularly traumatic events. But this process of awareness and recognition often needs to be done explicitly to let the emotion run its course. It needs space to be felt. Allowing for the emotion to pass through our systems, as Brach says, "like a visitor," can allow for it to be released. Ignoring, judging, or fighting the emotional event as it comes up in our bodies is often counterproductive. Allowing for the emotion to be felt, with no story, explanation, or judgment attached can often have profound results.

When we are exposed to other people's emotions, and they in turn trigger us, it can be helpful to adhere to these same protocols. People need to feel things in order to process them. Part of overcoming a trauma is letting it be felt and expressed in the right setting. If it's done in the wrong setting, it can trigger others and make things worse.

But no matter who you are or what you've experienced, bad things don't last forever. Eventually the wounds in society and in life will heal, and people will move on. No matter the depth of the hardship, things eventually get better given enough time. Whole societies do move on from shared tragedies, and it's possible for individuals to do so as well.

Social media, unfortunately, makes this process of moving on much harder.

Processing Trauma with Humor

Let's revisit comedy. At its best, comedy is a way of helping people process difficult things. Caring about something is part of the tension that comedy likes to skirt. We laugh because we care, or we know others that care. We "make fun" out of things that were not fun. It's a form of processing that we do together. A way of releasing tension and turning tragedy into laughter. But comedy fails if it's too close to the tragedy it's addressing—we sense that as being in bad taste, or "too soon."

In this way, comedy is sort of a democratic process for a viewing audience. If enough people think a bad joke is funny, it lands, and some portion of the audience is pulled along. If the joke is too close to a sensitive topic, people groan or, worse, they get angry (see Chris Rock and Will Smith at the 2022 Oscars).

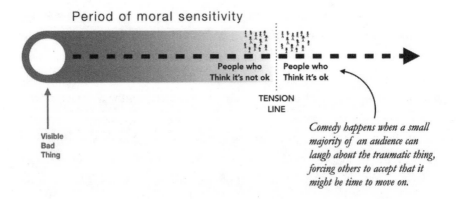

For example, in 2001, just a few weeks after 9/11, comedian Gilbert Gottfried made several jokes about flying planes into Manhattan at the beginning of a show. These jokes didn't land, and his audience booed him, calling out the words "too soon!" It wasn't okay to make that joke in the time immediately afterward. It was too close.[9]

But the opposite is also true. It can be "too far" from a topic for a joke to land. It's fine to joke about the Byzantine Empire's collapse, because no one alive today really knows what it was like when that happened. Back then, it would have been terrible to joke about. But today that old traumatic event makes a largely useless joke because no one has any stake in it. We're simply too far away, and no one cares.

When Everything Is Too Soon

As we're made aware of more *perceived* traumatic events, we find ourselves in a period of many increased moral sensitivities. With a growing awareness that so many things are traumatic, each of these topics becomes less humorous, less okay, and less acceptable to discuss without acknowledgment of the trauma.

The news is full of terrible events. Our feeds are full of people *reacting* to these terrible events. The more we're exposed to these things, the more suffering we see in our feeds, the more we update our moral sensitivities to adapt.

After enough time and enough exposure, many of us begin to see thematic relationships between these terrible events—signals where there may or may not be any. As our minds fill with violations, we see more potential dangers. Each thematic linkage becomes a new narrative—a sort of cultural myth or story that is shared among people with the same moral worldview. This process can be a potent grist for people prone to believing (and creating) conspiracies.

Taken together, these create a wide-open cultural minefield for us to collectively navigate. Each land mine triggers a new debris field and causes people to fear re-triggering the trauma. Each one updates the moral matrix in a slight, but measurable way. Each transgression, cancellation, and callout is a new type of update to what's considered "acceptable" in public life.

We are constantly looking outward from this hazy moral matrix of sensitivities. We sit at the locus of "now" in the middle of this current moral moment. It affects our vision looking into the future and looking into the past. If you wonder why, when you watch a movie from a decade or more ago, it may feel shockingly insensitive or problematic, this is why.

Our observations about the past and the future happen through
the lens of current moral sensitivities, dramatically influencing our
judgment of people, content, events and ideas

There are a few key things to remember here:

- Traumatic events are a difficult but expected part of living.
- All tragedies and traumas eventually fade. As individuals, we process them slowly. Society doesn't permanently remember them.
- The more traumas we are exposed to, the more sensitivities we collectively have.
- The more sensitivities people have, the more opportunities there are to hurt people's feelings unintentionally by triggering a cultural land mine.

- Social media has given status, power, and prestige to those willing to trigger these land mines beneath the feet of others, which can cause further visible traumas.

These are important concepts for recognizing that we are, more than ever, living in a time of profound moral sensitivities. It explains, among other things, why "cancellations" seem to happen on a regular basis and why comedians have been the focus of so many of these detonations.

Real Land Mines

When I lived in Cambodia, midway through the reservoir design process, we spent many days around the project site doing surveys by motorbike in the sweltering sun. We had to map the catchment area of the reservoir—an expanse of dry rice paddies that stretched to the horizon.

During one of these long survey expeditions, we sat with a community member under the shade of a coconut palm avoiding the hottest part of the day, drinking copious amounts of water. An old man named Mit Sen told us the history of the reservoir: exactly when it had broken, where the primary irrigation canals were, and when the worst of the fighting had been during the war. Learning about the battles that raged through the region in the '80s and '90s, I asked him about land mines. During my first visit to the reservoir, the monks had assured me that the area had been cleared of mines—fully safe to walk through and around.

"Oh no. We found two mines last month. A farmer lost an arm while harvesting rice—right over there." He waved his arm at an area we had just been walking through. As my Cambodian staff translated these words, our shock must have been apparent. "Oh yes. A bit dangerous," he said.

We had just been spending whole weeks trudging around fields that still had active land mines and unexploded ordnance.

The psychological change that occurs after realizing you've been walking through a minefield is something hard to put into words. Your vision and focus shift. Suddenly, the space you inhabit is unsafe. The pathway narrows, and every step matters. Your carefree attitude of exploration and curiosity becomes one of extreme caution and concern: Stay on the well-trodden trail. Do not deviate. Do not mess around. Just *learning* that someone nearby had been injured by a land mine can change your perspective on the possible paths ahead.

Navigating the metaphorical land mines of social media is not so emotionally different. Public cancellations of well-intentioned individuals often have real proximal effects. Seeing someone run afoul of a mob of angry people, despite their motives, changes what kind of creative and intellectual risks you might take. This causes a sort of constricting effect of speech and behavior, where people don't know what to say, or how to say it.

Let me be clear: Twitter is not a place where people lose limbs. Social media cancellation rarely has violent consequences, and the casualties of war do not at all compare to people's hurt feelings and damaged reputations.

But the effect is not negligible. Before you dismiss out of hand the impact of regular public cancellations, let's unpack how it can backfire.

A Million Dirty Looks

Social media is built on the back of many innate human impulses. The fundamental desire for prestige and social status underpins our desire for followers and likes, for example. Cancellations and callouts are a function of the human impulse to shame others when they transgress the moral norms of society. It's an amplification of a core human social behavior: gossip.

Gossip, as much as it tends to get a bad rap, is an adaptive advantage for humans. Gossip can act as a system of social communication, helping others indirectly learn about the world using other people's experiences and determine right from wrong. Controlled studies show that gossip can actually *increase* trust and social cohesion in public-goods games in which players need to know what people's reputations are. Gossip is often more effective and efficient than costly punishments to promote and maintain cooperation in social groups. Research shows that groups of people that utilize some amount of gossip are often actually better off[10]

Gossip amplified by social media, however, becomes something else entirely. A tiny act on social media can easily shame a loathsome individual.

Liking someone's funny post that humiliates your enemy feels good. Clicking on an article that explains the blunder, faux pas, or idiocy that led to an individual's downfall doesn't seem so bad, either. Indulging in collective schadenfreude is an oddly pleasant experience: a small retweet, a like, a share, a repost...it all seems innocuous, simple joy. A tiny guilty pleasure that brightens your day.

But each of these is a tiny vote for condemnation of the recipient. And each of these votes can tally up to a life-changing stream of admonishment for the target. Having the passive ire of one million people directed at you is a debilitating sensation, one in which your life can grind to a halt. Even a modest handful of negative news stories written about you is likely to narrow professional prospects, as your name becomes linked to every news article in the public record, one Google search away.

Of course, many people do deserve admonishment for bad behavior. Some views and behaviors are repugnant and should be railed against. For some people who publicly do or say horrible things, this can be a well-deserved callout and force them to change their behavior without a formal sanction. People are not being put in jail or physically harmed, after all; they are just being publicly humiliated for bad behavior.

Beneath cancellations and callouts is an attempt at unifying the moral frame that others are using. It's a way of saying, "This person is wrong." This feels good when we publicly signal to our communities that a line has been drawn around unacceptable behavior.

But like many events we observe on social media, this is an illusion of solidarity. The line is often not drawn as clearly as we think. If you are canceled in a partisan fashion, it can be a strange sort of boon. When people empathize with your reason for cancellation—after being outraged by the outrage against you—they may now want to follow you, support you, and use you as their own example. For those individuals who want to do battle in the so-called culture wars, the rewards can be sizable. A public cancellation can open up interview requests on Fox News or MSNBC and capture thousands of new followers for you, building a platform for speaking and writing. While you may become a pariah for one group, you might become a martyr for another. Some politicians often purposely skirt this line, and directly wade into areas of cultural sensitivity because they know it's a signal to their voters—a way to get them to pay attention. They want the outraged media coverage because they know it can raise their profile.

But for every person who's able to build a career atop the backlash to their cancellation, there are many more who simply retreat in shame, forced to pick up the pieces of their exploded lives quietly. When these detonations happen for unintentional gaffes or public misunderstandings—and they regularly do—these cascades of judgment can spill into their lives for reasons they don't necessarily understand, where they become a new defining life event.

To be clear—cancellation isn't just a left-wing or right-wing issue. Though studies show that cancellation is more often used as a progressive weapon, it's used by people on the right as well.[11] Attempts at cancellation happen to conservatives and liberals alike, and by conservatives and liberals alike. The threats and incentives of cancellation are becoming a new feature of life in a world dominated by social media.

What It Does to Everyone Else

Imagine playing a game of basketball in which after each pass, the rules are changed by the referee. At every throw, the players are forced to pause, and slow down to try and figure out what to do next. Can they still dribble? Can they still shoot three-pointers? Each pass becomes a point of friction and confusion as they reconfigure their strategy to get back to scoring. This game would be anxious and confusing for some, and profoundly infuriating for others. It would be hard to play for everyone.

Living in a society in a constant state of moral flux feels much the same way. The general anxiety it produces can be explained by one of the oldest sociological terms: *anomie*. Coined by Émile Durkheim in the 1890s, anomie describes a society whose norms are ill-defined and constantly changing. When there is not a common understanding of right and wrong, people have more difficulty moving cleanly in the space, cooperating, and generally getting along.[12]

A recent national survey by the Foundation for Individual Rights and

Expression (FIRE) found that fear of public callouts and cancellations was a widespread concern for a large fraction of the population.

> Nearly one-quarter of people surveyed are "fairly often" or "very often" afraid to state certain opinions for fear of losing their jobs or their standings in school, while 18% are similarly "afraid to say what [they] believe" for fear of the potential consequences. Eight percent admit they often feel pressure to say things they don't believe "in order to fit in."[13]

College students seem to have it worse. According to FIRE, 60 percent of students report that they've felt unable to express an opinion because of fear of reprisals from other people on campus. In an educational environment, this can be especially damaging to the exploration of difficult concepts and ideas that lead to a better understanding of multiple viewpoints.[14]

Life is complex and messy. Having opportunities to update beliefs is part of being human. As these tools become powerful cultural forces, we're finding ourselves pulled into a chasm of moral certainties with few gray areas between.

When there is a severe fear of public shaming, people often retreat from public exploration of ideas. In real life, public explorations of moral gray areas are often opportunities for learning and valuable conversation. When people hide their views away, they don't necessarily change their opinions; they may instead lead them further into tighter ideological groups—online and off—where their beliefs can stay safe and unquestioned.

Net Good or Net Bad?

There are plenty of horrific things that people should be called out for publicly. Not every quiet hesitation to share something online is bad. Not every public criticism is wrong. The trumping up of moral talk and attitudes clearly has influenced our public conversations. So how can we measure what's good and what's bad?

Minorities of all kinds—both ideological and cultural—often have to carry the responsibility of educating and updating the beliefs of people in a majority. This is tough work, and can be exhausting for individuals to do on a regular basis. Not having to regularly deal with offensive public narratives may feel like a helpful by-product of cancellation.

Moreover, some of the biggest, most significant social changes in recent years have come from activism catalyzed by moral outrage shared through social media. Many of these cultural and political movements wouldn't have been possible without them, and part of that is calling out power asymmetries and unaddressed grievances. #ArabSpring, #TeaParty, #BlackLivesMatter, and #MeToo all had some component of sharing moral violations in a public forum and calling people out for their transgressions.

To make sense of this issue, we might break it into two separate components. Social media has done two things simultaneously.

Increased *awareness* of moral transgressions—improved capacity to see the terrible things that are happening in society writ large. This, on its surface, seems like potentially a good thing. If we know more about what is wrong, we can work to make things better.

But, social media also does something else. **It has increased our *sensitivity* to moral transgressions.** It has heightened the number of possible perceived moral issues we face.

Later, we will try to answer the critical question: Which transgressions are valid? This is notoriously tricky. Perceptions of harm are fickle and subjective, going far beyond the legal definition of harm. If we shift the whole baseline of possible opportunities to make people hurt, and mad, and offended—and if we *pay people* to find more offense—then we're going to plumb the depths of historical grievances as far as they can be plumbed.

Recognizing just how powerful and problematic cancellations can be, we might then orient around a certain type of moral transgression and outrage that should be fought against at all costs: falsehoods, blatant untruths, and lies. These distortions, whether intentional (as is the case with disinformation) or unintentional (as is the case with misinformation) are shockingly pernicious, and deeply problematic for everyone exposed to them.

If you're beginning to wonder about ways of improving the system, we can start there: with lies, falsehoods, and mistruths. There is a thicket of problems to deal with, but they are worth addressing clearly, which we will do soon.

IN SUM

Outrage on the internet isn't always wrong. Public criticism plays an important role in making sense of our world. Many outrages that ripple across the internet stem from legitimate grievances and traumas.

Cancellations and call-outs are a function of the human tendency to shame others when they transgress the moral norms of society. It's an extension of a core human social behavior: gossip. Norms around gossip, studies have shown, can often be good for group cohesion.

But social media has created a system of amplifying our perceptions of traumatic events, giving them a much longer shelf life and much broader reach. Issues that historically might be processed between just a few people can quickly cascade to encompass millions of viewers, who all bear witness to the trauma. This updates the collective norms of what is appropriate to say, sometimes shutting down critical conversations that should be had, and dramatically increases opportunities for misunderstanding.

In light of social media's ability to make us aware of so many more problematic events, we seem to find ourselves in a period of rapid moral updating. There's no shortage of fresh opinions about what's right and wrong, acceptable and unacceptable. Next chapter, we'll try to answer the question: Where is this all going?

Chapter 14

The Waves of Moral Norms

Social media is rapidly updating our public notions of right and wrong, making the norms of the past seem very strange very suddenly. As a society, how might we expect our values to change in the coming years? Does the moral arc of the universe, as Martin Luther King put it, naturally "bend toward justice"? Or does it just bend in random directions?

We're all trapped in a moment. We were born into a time period that we had no part in deciding. The same way we don't choose our parents, we don't choose when we're born.

We're locked into a life that started at a particular moment, embedded in a set of norms and experiences that constitute "now." We're plopped into a moment, placed on a pathway with a bunch of other people traveling forward together. We have overlap with our parents, who traveled on this pathway before us, and our grandparents, who traveled before them.

This particular "now" might be worse than other ones. Looking back in time we can see that a lot of previous "nows" were terrible moments to be alive. Being born into a time of war. Being born into a period without running water. Being born in a time when the color of your skin legally restricted your rights.

For the most part, the most recent "now" is the best now, because humans get better at doing things, like making tools, laws, and rules that keep us happy and alive longer.

But now has constraints. Now has rules. Much of the time these are explicit rules, like don't urinate on the street in front of other people (a behavior that was fine in a different now).

But a lot of other rules are unwritten and are just there because they feel bad to break. These rules are based on what is acceptable to this particular now. These rules are based on the moral matrix of your community. When we reckon with the moral norms of the past, we're judging with the lens of the present.

So how do morals change over time? How does society shift its moral matrix?

Moral norms are strange things. They emerge from consensus in society, from communities inhabiting the same space and time. They change over long periods and in great waves. These changes come as a collective reckoning and realization, borne from tragedies made visible. Interpretations of sacred texts inform powerful orators, great thinkers, and eloquent writers, who then try to apply it to the moment. They make a case, people hear their words, and the concept of right and wrong slowly shifts. They're as much a function of deep moral values as they are functions of economies of surplus, and technologies that allow for more humane options.

As individuals, we usually ride these waves. We look to others to validate our feelings of right and wrong, and norms emerge. Cultural norms are established and values become standards, codified into laws.

Hegel: Society Moves Forward in a Dialectical Way

The philosopher Georg Wilhelm Friedrich Hegel is credited with developing a framework for understanding how collective truths are found in society. He did not think it was linear. He instead believed that views form from what can be called a *dialectic*.

The dialectic has a three-step process: the thesis, the antithesis, and the synthesis. He held that, in general, the best way to find truth was by placing two opposing ideas against one another. Beginning with a thesis—a proposition of truth—that must be challenged by an antithesis, or opposing idea. Only after these two concepts are pitted against one another does a synthesis emerge—the closest version of a truth that is possible at that moment. "Truth is found neither in the thesis nor the antithesis, but in an emergent synthesis which reconciles the two," he wrote.[1] Looking from above, the scientific world operates much the same way. Hypotheses are tested through experimentation and the results are put into the world. These results are refuted by other evidence, and new hypotheses are offered, synthesizing the old evidence.

Hegel believed that this happened with whole societies, too. Social progress is never linear, but swings from partial truth to partial truth. Lurching from one extreme to the other, we learn to figure out what is accurate and what is not.

A bit like a group of humans lost in the dark walking blindly. Someone declares a belief: "West is the right direction." (The thesis.) After walking in the darkness, they reach a deadly cliff, losing people over the edge. They have gone too far. Someone else, criticizing the decision, offers an alternative: "You see! You are wrong; the path is north." (The antithesis.) The troupe then travels north until, after a while, they reach the sheer face of the mountain, where they are stuck. "Ah," someone says, "the way is actually northwest." (The synthesis.) With each realization, they navigate from extreme to extreme, finding their way through the darkness.

In much the same way, society zigzags between moral extremes.

Hegel used the example of the French Revolution, in which the exuberant rush to democratic liberty was followed by profound chaos of the Reign of Terror, in which thousands were murdered. This turbulent period was then followed by authoritarian Napoleonic rule.

Similarly, in the United States, the postwar ultraconservatism of the '50s was followed by the activist extremes of the '60s and '70s, then manifesting in the liberal materialism of the '80s.

These swings of collective social opinion resemble something like a pendulum on a mechanized track moving forward in time, swinging from partial truth to partial truth as society figures out what actually works. In retrospect, these swings seem clearly reactive, but while we're in them, they can feel catastrophic and terrifying, or utopian and euphoric, depending on the moral matrix you live within.

But this is not just related to after-the-fact evaluations of social norms. It's much more applicable when we think about the tools and technologies we use and regulate.

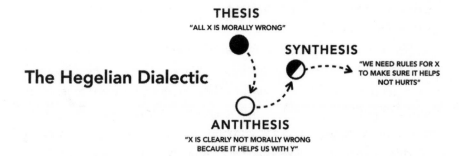

THESIS

"ALL X IS MORALLY WRONG"

SYNTHESIS

"WE NEED RULES FOR X
TO MAKE SURE IT HELPS
NOT HURTS"

The Hegelian Dialectic

ANTITHESIS

"X IS CLEARLY NOT MORALLY WRONG
BECAUSE IT HELPS US WITH Y"

Just as we're locked in time, we are to some degree locked into the dialectic of this moment. We're forced to adapt to the righteousness of our communities. We adapt to the moral matrix that we live and exist within. It's part of being human. It's part of living in society. We bow to the righteousness of now.

We don't like to think about how we're stuck within the limitations of this moment's dominant moral matrix. A strange, enmeshed space, enabled by the technological capacity and ethical standards of the present.

We Will Be Condemned by Our Grandchildren

Late one night when I was twenty-one years old, on a back road near the coast, I got into my first car accident. It was late, dark, and I was driving to a birthday party deep in the woods when a deer darted out in front of my car and clipped my front left headlight. I swerved after the fact, with no time to avoid it. When I pulled over, I found the animal maimed and struggling on the side of the road. It was a calf, not full grown but still twice the size of a large dog. Its neck was broken and it lay on the ground, intermittently flailing and trying to stand. After calling the local animal shelter, the county cops, anyone, there was nothing to be done. It was too late, and we were too far from town.

I sat with the deer for an hour in the dark in the woods by the side of the road, hoping it would die. When it didn't, I made the eventual decision to put it out of its misery. After retrieving an old baseball bat from the trunk,

I struck it in the head until it was dead. It was a brutal, honest, and tragic act that I'll never forget.

That moment changed my relationship with eating animals. After that point, I began to comprehend that I was actively outsourcing my discomfort to another person every time I ate an animal product. I decided to become a vegetarian until I could, myself, kill an animal I wanted to eat. Throughout my twenties, I embarked on a lengthy philosophical experiment as a result. One that brought me to a yearslong education, working from species to species, from fish, to chicken, to pig, learning as much as I could about the process of slaughter and consumption. Slowly I added new animals to my diet after learning what it took to kill them, clean them, cook them, and eat them myself. Today, I still try to eat only animals of a species that I've killed personally.

It also brought me to a broader understanding of one of the most painful outsourced human industries: large-scale factory farming. Allow me for a moment to step upon a soapbox, recognizing that this moral view is not shared by all. Bear with me, as I promise it has a point.

In 2020, 74 billion animals were slaughtered for food. That's roughly ten animals for every human on earth. Most of them live their entire lives in cramped, dark conditions. Many are bred to maximize weight gain and are aggressively fed a regimen of antibiotics and hormones to keep them alive through their brief and painful lives.[2]

I suspect that images from industrial meat facilities cause most people to feel some sense of revulsion. If you eat meat, you've consumed these animals. You've paid someone else to go work at one of these places, and supported an assembly line of animal executions.

An omnivorous species like ours has a moral conundrum built into our genes. We can empathize with our food; we can have compassion for these animals. Meat is tasty and good for our health. It just so happens to be attached to a living creature that can suffer. We're abstracted from this truth by a market economy that doesn't find much value in emphasizing its horrors. As Ralph Waldo Emerson said, "You have just dined, and however scrupulously the slaughterhouse is concealed in the graceful distance of miles, there is complicity."[3]

As someone who cares about this cause deeply, I was forced to ask myself: What might change this practice? How might we at least reconcile with it? People that work on these farms recognize these dilemmas, and many try to mitigate the suffering of these animals. There are many ranchers, researchers, and policymakers who do their very best to narrow the amount of suffering the animals endure through options like cattle runs and cage-free environments. Regulations do make a difference, and by and large the industry is getting better at reducing its worst practices.

But for every animal raised humanely, there are many more who live short and terrible lives, particularly in countries where few standards exist. When we examine this problem carefully, we're thrust into a complex dilemma. We *can* survive without meat. It's not survival that propagates this industry—it's comfort and luxury.

In India, for example, people make different choices. Four in ten people consider themselves vegetarians, and eight in ten limit meat in their diets. Dietary laws and traditions are part of India's most widely practiced religions. In Hindu texts, vegetarianism is often praised, and cows are traditionally viewed as sacred, so many Hindus may refrain from eating beef entirely. This is one of the largest countries on earth, and they manage to consume less meat than the rest of us.

But this isn't a book about your moral obligation to not eat meat. For me, a meat eater, it illustrates how embedded we are in our place and time. It reminds me that we're stuck in a narrow band of society-wide complicity in a very hard thing. And this hard thing, an interspecies industrial holocaust, will likely be looked back upon by future generations with real horror. It's a stark reminder for me that we, as a society, will likely be morally condemned by our grandchildren. Some of our most innocuous customs may be looked back upon with horror and fury by future generations.

Social norms become entrenched inside of economic systems, and because of this, they become difficult to change. Many of our beliefs and our behaviors today will not survive the moral reckonings of tomorrow.

Beyond large religious movements and cultural norms with centuries of precedent, what might cause moral changes in society? Outrage over the injustices of the moment *do* result in behavior change. They advance laws that cause industries to evolve. They create demand for better options in the market, sometimes with real, positive results. Opportunities arise from these sentiments. Technology, as much as it pushes us in profound, unforeseen directions, can also make moral problems fade away. Economies that adapt to new morals can sometimes make things much better.

In the late'70s, for example, enormous media attention began to focus on the rapidly growing ozone hole near the planet's poles. This created a broad concern about chlorofluorocarbons, or CFCs, chemicals used in refrigeration and aerosols that are especially harmful to the ozone layer. Ozone depletion was on track to inflict havoc on the environment and dramatically increase UV radiation in certain parts of the world, as well as increasing temperatures. It was one of the first major environmental movements to capture national attention, and it was surprisingly nonpartisan.

Governments and industries debated for years about what to do. After there was a consensus, supported by evidence, as to the problem at hand, people began to look for solutions. Those solutions came in the form of several alternate technologies that happened to be easy to implement. HFCs, or hydrofluorocarbons, could easily take the place of CFCs and, while still a potent greenhouse gas, were far less harmful to the ozone. A 1987 worldwide ban on ozone-depleting chemicals prevented a catastrophic increase in damage to the ozone layer, largely because there were simple alternatives. Since then, a number of even less harmful technologies have been identified, with most nations beginning to ban HFCs in favor of even greener alternatives. In 2022, measurements of the ozone hole over the South Pole showed it at its smallest since the ban had taken effect.[4]

The reality is that most problematic behaviors change en masse only when it becomes easier for people to do so. Outrage can catalyze these changes, but oftentimes it's also just someone building a better option and offering it to consumers.

Many moral problems become solvable *because* there are technological alternatives. Our grandkids' kids will likely see technologies available in their lifetime that will give them a fundamentally different perspective on the moral choices available to them. They will likely have meat that was not farmed inhumanely. And if history is any guide, they will judge us the same way we judge some of the ridiculous norms of our great-grandparents' generation. They won't have to make a hard choice when they decide to eat meat. There will simply be better options available to them. Their conscience will be clean due to the hard work happening now.

A fun way to explore the oddities of these embedded moral norms is to search for vintage ads on the internet. If you peruse advertisements from your great-grandparents' generation, you'll likely see some really strange stuff. Doctors telling you smoking is actually *good* for you. (More doctors smoke Camels!) An ad that implies that beating your wife is kind of OK if she doesn't get you the right coffee. An ad showing a young girl playing in bed with an "absolutely safe" Iver Johnson revolver. ("Papa says it won't hurt us.") We look back upon these ads with shock, revulsion, and maybe a hefty dose of amusement at their backwardness. These are tiny portholes to a time with a different moral matrix.[5]

Source: R.J. Reynolds Tobacco Company Advertisement, More doctors smoke Camels than any other cigarette, 1946.

Source: Chase and Sanborn Coffee Advertisement in Life Magazine, If Your Husband Ever Finds Out You're Not "Store-Testing" For Fresher Coffee, 1952.

Source: The Canadian Magazine, Iver Johnson Revolvers, 1904.

To me, these ads help frame the extremity of our moral judgment at any given moment in time. It doesn't at all excuse the mistakes of previous generations, but it reminds me to be gentle with our moral condemnation of each other in this current time. Great moral progress is made easier with better tools and innovation, just as much as laws that help us avoid the difficult moral trade-offs of the past. We're practically guaranteed, particularly in this moment of righteousness, to be quizzically studied, judged, and condemned by our descendants for our strange behavior.

Language Changes Faster Than Morals

Imagine waking up one day and learning that everyone you know below a certain age has contracted a mild flu overnight. The flu did two things simultaneously. First, it updated the language-processing center of the young person's brain, and made them all selectively perceive that the word *cat* is now pronounced "taco."

Second, it also affected their memory, removing any recollection of those particular words ever meaning anything else. *Taco* now means "cat" to every young person you know. It was never different. Strangely, only people above a certain age—those you think of as stodgy and out of touch—can still see and pronounce "cat" with no impairment, and speak the word perfectly. Now, when you say "cat" to your young friends, they think you sound like an old fogey and judge you for it accordingly.

What would you do? Would you think you were a little crazy, and go to the doctor to get your ears checked? Or would you just laugh it off, update your language, and play along?

The above scenario is not dissimilar to the feeling many people have nowadays when they try to communicate with those who have grown up on social media. As social media has radically increased the speed at which we consume information, it has also increased the rate at which our language is evolving.

In the world of physics, the term *half-life* is the time required for a quantity of something to reduce to half of its initial value. It's used as a measure of how quickly atoms decay by half, or how long before chemicals in our bloodstream are reduced by half. Caffeine, for instance, has a half-life of about five hours. If you have a cup of coffee at twelve p.m., then by five p.m. half of it will still be in your system. The term can be used to describe the decay of the usefulness of many things.

The language we speak has a half-life, too. Looking backward in time, English becomes less and less decipherable the farther you go. In the 1400s, you would call a winged feathered creature a *b-r-i-d*, and you would spell it that way in English. This changed when someone misspoke *brid* and it became, after a time, a *bird*. (It's easy to imagine children of that era being scolded for calling it what their descendants later would consider correct.) Even the US Constitution uses an "incorrect" spelling of the word *choose* written in it, as *chuse*.

In that same decade, Alexander Hamilton, one of the most famous orators and writers of his era, wrote this in the *Federalist Papers*, in an attempt to speak clearly for the public to understand: "An enlightened zeal for the energy and efficiency of government will be stigmatized as the offspring of a temper fond of despotic power and hostile to the principles of liberty."[6]

This is American English written just over two hundred years ago. To your average English speaker today, it's a challenging sentence. It makes sense, but you need to read it several times for it to make sense. (You might imagine Hamilton alive today, reading Twitter, and being similarly baffled by an average tweet about any modern hot-button issue.)

The accepted norms of any language will change, through a process of misinterpretation, innovation, and natural evolution. Language simply morphs over time. The terms we use to describe important things in society—especially politically contentious ones—change even more rapidly. This is called *semantic drift*.

Words related to our identities change especially quickly. One of the most well-known organizations fighting for the rights of African Americans is the NAACP, the National Association for the Advancement of Colored People. Founded in 1908 by Dr. W. E. B. Du Bois, Mary White Ovington, Moorfield Storey, and the abolitionist Ida B. Wells, the NAACP used "colored" as an inclusive and progressive term at that moment in time. The term PoC, or People of Color, has come back into common use recently, but it's considered pejorative to call someone a "colored person" today. The actor Benedict Cumberbatch received criticism when he used the term "colored actors" in a 2015 interview in which he lamented the lack of roles for Black actors in the UK's film industry. His sentiment was explicit: We need more roles for Black actors, and it needs to change. But when he referred to them as "colored actors," he indicated that he wasn't aligned with the social norms of people close to the cause.[7]

The language we use is a critical signal for determining where we exist on the spectrum of accepted social progress within our in-group. Our vernacular is a sign of how "with it" we are and how attuned we are to our in-group's current moral norms.

The method by which language changes is usually defined by subcultures and youth, and exposure to evolving attitudes on contentious topics. It's also a function of exposure to new ideas and conversations. Over the last century, largely due to the consolidation of media monopolies, these types of linguistic changes probably happened more symmetrically. Watching the same TV channels and listening to the same radio shows might expose kids and adults to new tropes and slang all at once. As the internet has blown apart our shared media, it has also blown apart our ability to easily track changes to our shared language.

But if acceptable language shifts over time, it still hurts when we are corrected for it. For many of us, linguistic norms are moving faster than our ability to catch up. Some people respond to this change in acceptable language with a feeling of anger. ("I could say this when I was younger! Why can't I say it now?") Beneath this anger is a type of sadness—a recognition that time is passing, and a fear of aging. A feeling that society is moving faster than one's ability to keep up with it. A loss of sacredness and a loss of relevance.

Staying relevant in society is a type of power. We want to know how to navigate the world and say things that still have influence, to know that we still have a voice that matters. For many people, losing that, or feeling like it's being discarded, is a painful loss of identity. It's society saying, in a not-so-subtle fashion "Your ways don't matter anymore."

This feeling of exclusion is not dissimilar to an identity threat. It often pushes people to find safety in other people who feel the same type of exclusion. Many political candidates are willing to capitalize on this anger by portraying these transitions as a fundamental breakdown of society. For many of us standing in the middle, this pressure to change is hard to reckon with, particularly when we're exposed to so many opinions about what is now unacceptable.

It's important to remember that we will all be caught in the moral transitions of acceptable language at some point. Eventually we'll all feel the sting. Youth defines the next iteration of language, and after enough time *everyone* loses touch with it.

Many aspects of life are moving faster than we can process. What's critical is assuming good intentions. In general, people *want* to connect and communicate. Most do not want to offend. What's crucial is recognizing that it's happening, and navigating the ends of the spectrum of change with compassion and good faith.

IN SUM

We began this chapter asking a few big hairy questions about changes in moral norms. We explored how we are trapped in the moral matrix of our time, and how moral norms shift in a process of zigzagging between extremes which Hegel called a dialectic.

The norms and customs of today are unlikely to persist forever. The same way we look back upon some of our great-grandparents' pastimes as backward, our great-grandchildren (or even grandchildren) will likely look upon many of our norms (of eating factory-farmed meat, for instance) with real horror. Understanding that these changes are inevitable doesn't excuse bad behavior of the past, but it should soften our condemnation of it, reminding us that society is constantly updating its values.

We learned that all language has a half-life of utility before it changes due to semantic drift, but social media is making us modify language faster than it has in the past, while also providing incentives for corrections and call-outs that are hard to stomach. Call-outs, done badly, feel like real threats to our identities, and encourage people to search for safety in ideological in-groups.

We've spent the last several chapters exploring the rapid shifts that social media has wrought upon our culture and society. From here, we'll shift gears and focus on how this matches an important hidden pattern behind every new major human invention.

Chapter 15

The Dark Valley

The Upgraded Road

When I first moved to Cambodia in the early 2000s, there wasn't a single stoplight in the country. The vast majority of roads were crumbling remnants from before the 1975 civil war or dirt thoroughfares constructed from a local red clay called laterite. These roads were generally rutted, full of muddy potholes and impossibly deep puddles. They were painfully slow to drive on; you simply couldn't drive fast without bottoming out your car or breaking your motorbike.

In the early 2000s, a joint UN and World Bank–funded program finished building National Highway 6, which was at that time the first and only major paved road in that part of the country. It stretched for many miles from the capital, all the way to the border with Thailand. It was a normal road by Western standards, two full lanes with a paved shoulder wide enough for cars to pass on either side. As far as roads are concerned, it was nothing special.

But for local motorists used to driving on mud, this new road was a superhighway. It was the first time a whole generation could drive faster than fifteen miles per hour, which they promptly did. People flocked to the road in droves. Suddenly able to commute to and from their villages easily, people would open up the throttles on their 200cc Honda motorbikes and fly down the roadway at an astounding 50 mph. The results for public safety were catastrophic.

When your top speed is fifteen miles per hour, you don't need a helmet or worry about high-speed collisions. You're throttled naturally by the road's inherent impediments, and an accident at low speed is not likely to kill anyone. But at 50 mph, a motorcyclist is very likely to kill themselves or someone else on impact. Sadly, this began to happen with shocking regularity.

Between 2001 and 2007, road fatalities tripled in the country.[1] A week didn't go by when I didn't pass a dead or dying motorist splayed out on the road. I watched several terrible accidents happen in real time. These tragedies—twisted wrecks, horrible road injuries, and head-on crashes—steadily became a staple of daily life. We called emergency services whenever we could, but still, we all began to become desensitized to the regular sight of an occasional body on the street. For those just emerging from the mud and sand roads of the countryside, along with the improvements to efficiency and speed, the upgrade was deadly.

This isn't dissimilar to what happened to our media infrastructure between the years 2009 and 2012, when we essentially went through an enormous upgrade to our communication thoroughfares, increasing their speed, spread, and reach.

The trite and now clichéd term "information superhighway," while not particularly apt when referencing the introduction of the World Wide Web, makes more sense now when applied to the extreme and sudden availability of high-velocity information that now defines the viral era.

This burgeoning age of high-speed emotional content created a whole new category of catastrophic collisions of opinions and emotional content, resulting in casualties of viral outrage that now dominate much of our culture. We were all much like rural farmers emerging onto a freshly paved urban roadway, initially blissfully unaware of the dangers of our freshly acquired speed and connectivity. And now several years later we find ourselves oddly desensitized to the carnage littering the roadway.

In more developed countries, of course, we spend much of our time hurtling along at 60 or 70 mph every day with few problems, the result of years of public education on driving safety and traffic enforcement. Though driving remains one of the most deadly activities we participate in daily, we've mitigated these effects by creating systems, laws, and norms that make sure people know what they're doing when they get behind the wheel. Regulations make cars and drivers as safe as possible. This was an incremental process that took decades, something that we might begin to think about as we watch the wreckage around us pile up.

The Dark Valley

As much as we'd like to assume that the terrible by-products of our technology are fully foreseeable, this is not always the case. This doesn't absolve responsibility for fixing them, but in the very beginning, many of these problems are inherently invisible. Few expected that when the first Model T rolled off the tracks, automobiles would be the source of a worldwide climate crisis a hundred years later. More recently, few early adopters would have expected that Bitcoin would ever reach the scale that it would contribute greenhouse emissions equivalent to a whole country.[2]

Social media is full of edge case harms that were largely invisible in their smaller incarnations. With every new technology, the negative outcomes of their usage at scale are profoundly difficult to foresee.

There are clear reasons why harms are overlooked. A company building a new product needs to prioritize where its limited resources go. Early on, these resources usually need to go toward growth and keeping their existing users as happy as possible. Every early product has bugs and edge cases that need to be eventually addressed, but at small scale, these issues really don't matter as much. For every one hundred users, if you have one who is dissatisfied because of a bug or a problem, an early-stage company would be unwise to allocate all its efforts to solving that single user's issue.

But once a company reaches a million users, these small problems become categorically huge. That "one percent of users" issue, while still proportional, is now a ten-thousand-person problem. It might be something simple, like a bug that makes the experience terrible for that small fraction of customers, or they might be problems for society writ large—when a tiny number of bad actors use the product in a disproportionate way to harm others. It can also be another type of externality that is simply a by-product of scale: like the way products enter the waste cycle and become pollution, affecting everyone else later on.

But almost every major new technology that reaches widespread use has an aggregate of unforeseen harms that can reduce human flourishing for a period of time. If we're lucky, these harms become better understood, interventions are made, and society begins to deal with the worst harms and integrate with the tech for maximum flourishing.

We can map this to a clever tool that evolutionary biologists use for mapping how species flourish, using a graph showing how organisms react and adapt to their environment. This is called a *fitness landscape*, a term used because the graph resembles a mountain range. In this mountain range, there are peaks, or maximums—points from which all paths go downhill to lower fitness. There are also valleys, or minimums, where paths go uphill. (Remember the algorithm ascending to the maximum point of attention capture? A similar graph works here, too.) Species ascend the mountains of fitness, to a greater maxima, when they find success in their environment, and are able to thrive and reproduce.[3]

While fitness landscapes are helpful for biologists determining how animals evolve different traits to improve their reproductive success, they are also a helpful metaphor for thinking about how humans adapt to new technologies.

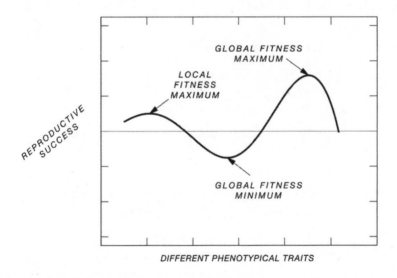

DIFFERENT PHENOTYPICAL TRAITS

We reach a local height of our "fitness" when we create a new piece of tech, but as the negative effects pile up with scale, we descend into a valley. Fitness, in this case, is an analog for collective human thriving. When widespread use causes huge new issues to emerge, we find ourselves descending deeper until we can find our way out, eventually climbing to a higher plateau. These valleys can be deep (when harms are extreme), or they can be shallow (when harms are light). Sometimes they lead higher (when we figure out how to integrate the tech), sometimes lower (when we don't).

But while this tech is being introduced, many problems are often hidden. They only become seen after a huge portion of society begins using them. In this way, they are invisible to the vast majority of us until it's too late. Eventually, this period of hidden externality—what I call *the dark valley of hidden harm*—is traversed after society begins to respond to the negative effects. If we graph this out, it looks like this:

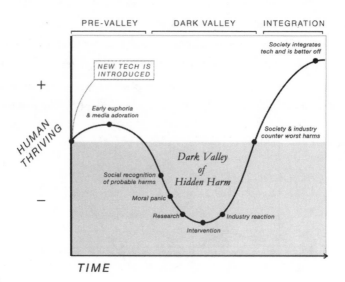

Most major technologies go through some version of this graph, in which there is invariably widespread early adoption and late social recognition of the harms.

What then occurs in the valley can happen in almost any order, but usually there is some combination of the following: recognition of societal harm, research on these harms, and a combination of study, panic, and politics that inform interventions, which may (or may not) help to reduce the harms of the tech.

The reason this valley is "dark" is that the harms of the tech are often hidden from view until mass adoption happens. A technology's adoption curve looks like this. New technology is introduced to media adoration and user euphoria, eventually reaching full adoption.

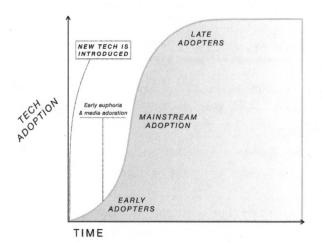

Adapted from Geoffrey A. Moore's book *Crossing the Chasm.*

If we were to overlay the adoption curve over the dark valley, you can see that it's visible only after it reaches the masses. It's usually obscured both by the early euphoria of new users, and by the sheer lack of visibility into the edge cases that might cause enormous problems later on.

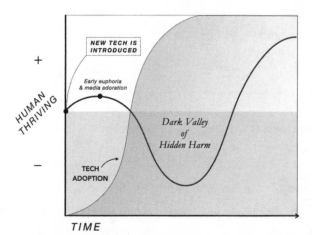

The dark valley of harm is "hidden" because widespread harm is often obscured by the flood of new users, or industry lobbying to keep a product in mainstream usage.

Over time, we collectively descend into then ascend out of these valleys as we struggle and adapt these new tools into our lives.

Using this frame we can break new technology into three stages:

- **Pre-Valley phase**—tools that have not yet reached widespread adoption.
- **Dark Valley phase**—tools that have reached mainstream usage, and are in some process of causing harm without a clear strategy or desire to fix them.
- **Integration phase**—tools that have begun to be integrated into society after addressing their most significant harms.

Each phase has a set of priorities and questions that can be answered and addressed to reduce the depth and duration of the valley itself.

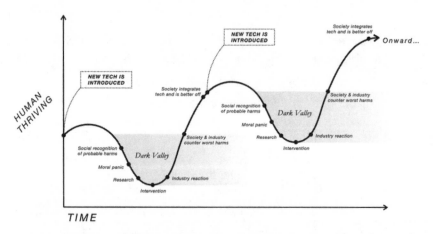

Humans adapt to new technologies like a fitness landscape, reaching higher (or lower) as we integrate and move between technologies.

Some technologies can have multiple valleys that are traversed at different speeds. The automobile, as we'll see, had huge problems beyond traffic deaths that needed to be addressed over decades, and still today has a whole new set of issues related to greenhouse emissions that we have yet to fully resolve.

Not every new bit of tech has a dark valley. Many of the inventions we encounter today come in the form of incremental improvements to existing tools that are deep inside industries: tiny changes in materials manufacturing that make metals or plastics stronger, or more cost-effective to produce. Improvements to waste-treatment plants that make them more sanitary and less likely to spread disease. Modernizing transmission substations in a power grid to more efficiently deploy electricity. Bits of code that upgrade the speed of a network. These smaller incremental changes do not tend to have serious system-wide impacts.

And some dark valleys are very benign, or simply aesthetic, such as a city's adoption of ride-sharing e-scooters, which tend to pile up in unsightly ways in metropolitan areas, until clear parking regulations are put in place.

But new *media* technology has an especially perilous valley when it comes into widespread use. When we begin using it en masse, it updates our capacity to *see* and *understand* what is happening in the world around us. That includes our ability to see and understand the harms of the technology itself. New media tech often creates moments of system-wide blindness and confusion, where both problems and solutions become murky and difficult to discern.

Think for a moment about how we are trying to determine what is "wrong" with social media. As we discussed, many academics and writers (myself included) were convinced that there was a clear ideological exposure problem, and that our perspectives were being summarily isolated by algorithmic sorting of opposing opinions.

But if you ask conservatives, the primary problem with social media is that there is a coordinated plot to algorithmically suppress voices and perspectives on the right. If you ask a liberal, you might hear about how these same algorithms *promote* conservative voices and hate speech. Each of these perspectives can be backed up with real-world examples (e.g., Trump being deplatformed, Trump building and amassing an enormous platform, respectively). How might we discern what the *primary* problem is in a moment when this tech becomes ubiquitous, and begins to dramatically influence our sensemaking systems as a whole?

Some dark valleys are shockingly tragic for humanity. Let's return to the automobile: In the 1920s, General Motors introduced tetraethyl lead as a standard additive to gasoline. It was initially known to be dangerous; GM began adding this lead-based compound to gas to help raise compression and performance in cars.[4] But the by-product was lead—a neurotoxin—being aerosolized into the environment in car exhaust. Lead is a devastatingly harmful chemical compound that causes cognitive disorders, aggression, anemia, measurable losses of IQ, and death.[5] It's been known to be highly toxic for hundreds of years. During early production of this additive, a number of autoworkers died from lead exposure. In the face of this bad news GM went on a public relations blitz, and prevailed in convincing the US government to allow for its compound to be added to gasoline, which soon became a global standard.[6]

It took more than fifty years to correct this terrible mistake. When additional research was done in the 1970s and 1980s, experts concluded that leaded gasoline had created a public health catastrophe that could be linked to increased crime rates around the world, trillions of dollars in economic damage, and millions of deaths. When the EPA then banned it, other countries began phasing it out. This particular technology, driven by profit, caused an enormous net reduction in human thriving. While there is now a strong scientific consensus that no amount of lead in the environment is healthy, this dark valley lasted more than fifty years and created one of the worst public health disasters in modern history.[7]

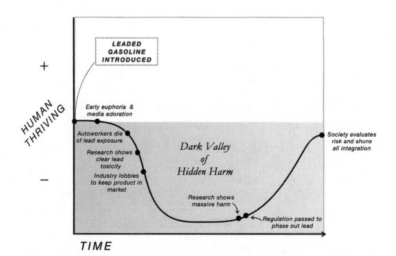

Integration Is No Guarantee

While most of our tools follow a path of integration, in which we become more aware of the issues the tools cause, and we subsequently manage most of their harms, sometimes we get it wrong.

Nuclear power, for instance, was initially regarded as a salve to all energy woes. But it languished after huge disasters like the Chernobyl meltdown, and perceived disasters (in retrospect, a moral panic) like the Three Mile Island partial meltdown, which soured public opinion about the viability and safety of nuclear energy. Today, their major harms have been largely addressed and nuclear energy remains one of the safest, most environmentally friendly technologies for energy production, even when accounting for its accidents.[8] The most significant critique of its adoption comes down to cost, as it is now cheaper to produce solar and wind energy in most areas of the world.

Most models for reaching emission reduction goals globally require some amount of nuclear energy. But as a result of the stigma surrounding the technology, it has remained politically off-limits. As a result of the fear of future disasters, nuclear power plants have been decommissioned and replaced by much dirtier forms of energy like coal plants. According to most experts, a Chernobyl-like event is simply not possible with modern reactors.[9]

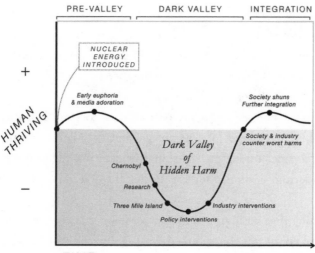

The worst path is when a technology becomes so widespread, and the harms become so substantial, that humanity doesn't escape. This is the fight we are in now—to find a way to healthily adapt to our new technologies fast enough for us to reach the next plateau.

Is social media leaded gas, or is it nuclear power? How bad is it for society? We must do an accurate accounting of its harms to understand where it has gone wrong, and where it has gone right.

Confusion Inside the Valley

Figuring out what's actually going wrong with social media has been inherently hard. Bear with me as we step back into the 2016 election one last time to explore what it feels like, in modern times, to try to make sense of things as we descend into a dark valley. The point when most of us seemed to wake up to the categorically bizarre effects of social media was the day Donald Trump was elected. This event was a fundamental shock to the established order and process of traditional politics. Pretty much everyone was confused. The Trump campaign itself even seemed surprised as to how they had pulled it off.

I attended college in a part of the country—deep red Pennsylvania— that contributed to Trump getting elected. As a result of my years in a red county, I had a modestly diverse political feed. In the weeks following the election, I watched dozens of my friends on Facebook unfriend one another in an unfamiliar show of a new, strange solidarity. A type of moral filter was being applied to old friends and acquaintances on the platform, a sort of purity test, essentially saying: *You're with us or against us.*

I saw plenty of self-righteous posts flow across my news feed, along with deeply felt messages of fear, anger, and notably, existential despair. On the other side, among my red-state friends, I saw reflections of joy, levity, gratitude, and optimism for the future. The division couldn't have been starker.

The thing that both groups had in common was very apparent: a sense of profound confusion about how the other side couldn't understand their perspective. It felt implausible that we would spend this much time cohabitating in a country and not know that Trump was a true contender. Even many of his gleefully irreverent supporters didn't consider him a real, viable candidate. The day he was elected the internet exploded with think pieces, espousing some configuration of the following:

"Clinton was just a poor candidate. The media was misinformed. Pollsters were wrong. Hillary was corrupt. Trump was a political mastermind playing 4D chess" ... etc.

What was really happening?

There was another idea, which seemed to be a plausible explanation for what was going on. Perhaps we were so entrenched in our ideological enclaves that we simply didn't have exposure to what others were seeing?

This concept was popularized by author and entrepreneur Eli Pariser, who had written the book *The Filter Bubble* just a few years prior. According to Pariser, what was happening in our politics and in our relationships was essentially a segmenting problem caused by algorithms.[10]

On the surface, the hypothesis seemed correct: The more choices we are given in our media environment, the more likely we are to isolate ourselves from the opinions of others. The more we're served stuff that confirms our biases, the more we're likely to believe our own dispositions. If Google and Facebook kept you ideologically segregated, how would you even know?

It also seemed to be backed up by an academic theory. In social psychology there's a framework called the *contact hypothesis*. It suggests that prejudice is reduced through extended contact with people who have different backgrounds, opinions, and cultures than ourselves. Developed by psychologist Gordon Allport as a way to understand discrimination, it's still widely cited as a successful way of reducing prejudice and increasing empathy—a time-tested way of helping people get along.[11]

The logic goes as follows: The more time you spend with others who are different from you in an environment that's mutually beneficial, the more you'll understand them. The more you understand them, the less prejudice and implicit bias you will have. More connection equals more empathy.

Illustration showing Eli Pariser's concept of the Filter Bubble.

What if most Democrats weren't spending enough time, virtually, with Trump voters? If filter bubbles were isolating us, then it was plausible to assume that liberals were mechanistically losing their opportunities for empathetic connection. In a world of news feeds, these algorithms would in essence store and serve information based on our preferences.

The result of this would be a stream of information that's entirely oriented toward fulfilling each individual's confirmation bias; a feed that's optimized for our attention will also be optimized for supporting and reinforcing our existing worldviews.

The hypothesis was plausible: Humans have strong confirmation biases and desire to consume information that agrees with their dispositions. We pick our books, our magazines, our movies, and our news stations. The internet has allowed for ever-more-fragmented consumption. Why wouldn't this happen in the online world, especially if these tools were designed for it?

Professor Chris Bail of Duke University decided to find out. With colleagues, he built an experiment in which he paid real humans to follow bots of the opposite political persuasion. The goal was to test whether people exposed to alternate political viewpoints would change their opinions—essentially testing whether breaking their filter bubble could make people dislike each other less.

They assumed that as people were exposed to differing political ideologies, they would orient themselves toward some type of more open consensus and understanding of the other side's perspective. The study measured their "affective" polarization beforehand and afterward—how much each person disliked the other side. They tested the assumption that bursting these filter bubbles would increase affinity and understanding of other political perspectives, and that polarization would go down.

What the researchers found was a shock. Among both groups, liberals' and conservatives' affective polarization actually went *up* after being exposed to alternate political viewpoints. It went up significantly more for Republicans than it did for Democrats, but both parties became less tolerant and more polarized. Both had even *worse* opinions about the other side after being exposed to different viewpoints.[12]

This went counter to many assumptions about what filter bubbles and echo chambers were doing to us. People weren't hiding in exclusively view-confirming bubbles. Something else was going on.

This view of what was happening to our conversations on digital platforms makes more sense when we remember what the types of content tend to go viral: extreme and inflammatory content.

On social media these users weren't exposed to the best version of other people's arguments. Instead, they were usually exposed to the most *extreme* perspectives. So, following a political bot of the other side was unlikely to be persuasive. Instead, these perspectives were far more likely to appeal to their political in-groups as opposed to more moderate views that might help sway someone toward their perspective.

Filter bubbles as Pariser described them don't really exist—something else more pernicious was happening. The moderate views, not the opposing ones, were lost when social media was in play.

As we descend into the dark valley of a new technology, making sense of what is actually transpiring is extremely hard. People have theories, but even the most plausible ones don't always work out. Pariser's hypothesis did, of course, help inform the conversation and the study of what might have been going wrong. He did greatly advance our thinking, even if the reality turned out to be surprisingly different.

IN SUM

We began this chapter learning about how societies often struggle to adapt to new technologies. In Cambodia, it was something as simple as a paved road that caused major problems for a generation that had no precedent for high-speed driving or traffic laws.

This maps to a consistent pattern we see with many new technologies: what we can call a dark valley of hidden harm—a period of negative side-effects that comes after widespread adoption.

We explored how this period of adaptation resembles what biologists call a fitness landscape, when species move from a local maxima downward before finding a higher level of evolutionary fitness.

We explored the three phases of this transition, *pre-valley*, *dark-valley* and *post-valley*. In these phases, early euphoria for adoption leads to some combination of social recognition, moral panic, and research that eventually informs interventions that begin our slow and steady climb to integration.

We also explored two major examples of this from the last century. First, tetraethyl leaded gas, which despite early research showing that it was harmful, took over fifty years to be remediated. It caused huge society-wide damage through its use, after which it was almost entirely banned. Second, nuclear power, which also had sizable harms, but ones we largely learned how to mitigate. We still avoid new nuclear power, largely because of its spectacu-larly vivid failures, even after managing its most significant risks.

A core idea of this chapter is that *most major new technologies are likely to have a dark valley of hidden harm.* Large-scale adoption of new tools will have externalities that negatively impact human thriving. Rather than trying to avoid all harm, we should instead learn to *anticipate* some bad outcomes, par-ticularly in times of media euphoria and rapid public adoption. If we expect these dark valleys, we can instead focus on decreasing their depth (how bad they are) and their length (how long they last) through a process of steady examination and reflection.

We also learned that new media technologies have an especially challenging dark valley, because they often influence our sensemaking ability—the process we collectively use to determine which threats are real. We explored one recent example with social media during the 2016 election, but we're about to examine many, many more.

For part III, we're about to take a journey back in time, exploring some of the most monumental media disruptions in history. Each will help us understand the consistent pattern of disruption and integration, and teach us how we might manage the harms of social media. This moment we're living through is not without precedent, and we've been here before.

PART III

History of the Machine

Chapter 16

The Ancient History of Virality

Our species has been around for a long time.

For most of human history, knowledge would just dissolve. Terrible atrocities, wars, massacres. Incredible heroism and profound triumph over struggle. Whole languages, cultures, and peoples could be subsumed by another. All was forgotten.

This knowledge would not stick, because there was no medium upon which to attach it. Names, faces, events, could be retold as oral histories, but they would not remain for long. Only some would transit forward in time for a handful of generations before they, too, were absorbed into myth, or lost forever.

Until we could write, we could not record. Until we could print, we could not compare.

Understanding the history of our species is like looking backward into a picture frame that holds tiny scraps of a long-faded image. We can make out the rough outline of human forms. We cannot see faces, distinct features. The image has dissolved, and we must infer and apply our own interpretations to their lives and struggles.

But what we do know is that across all cultures, everywhere on earth, humans have a desire to learn and share news. Our ancestors lived in an information-poor world, and had few ways of exposing themselves to new knowledge. Because of this, our brains might have evolved to crave novel information in much the same way we crave sugar. Both were helpful for our survival, and both were rare in nature.

For much of our past, up-to-date knowledge was also a matter of life and death. It helped us understand where danger was, whom to trust, and what to avoid. The news was a ritual of exchange, not just a commodity to consume. The origin of the word *news* is quite literal: It's the plural of the "new" thing(s) people know about and pass along.[1]

The ancient world was smaller not just in terms of populace, with a global population a tiny fraction of what it would become in the Industrial Era (most of the largest cities on earth had far less than one hundred thousand people), it was also smaller in terms of perspective.[2] Most of the world was basically invisible—a vision incomplete and fleeting. Life was slow, cyclical, tedious, and local. Any news was a glimpse of the unknown.

In rural areas, when a traveler carrying news arrived, it might be the most significant informational event of the year. A piece of written parchment or a letter, read aloud in public, was a new window into a foreign world, an alternate reality. News of major events traveled shockingly slowly. Like light emitted from a far-off star, the view of an occurrence was only known long after it occurred.

The story of civilization is the story of slowly evolving news networks. Each network was painstakingly built and improved over centuries, influenced by every fledgling technology that improved the transit of knowledge.

This knowledge was about power. The faster and farther news could spread through these networks, the easier it was for rulers to rule. These networks were so important to those in power that empires would spend lavishly to ensure their smooth operation. Records from the Yuan Dynasty during the reign of Kublai Khan show that his postal service included more than 1,500 postal stations stretching across the continent, employing couriers to rush deliveries across his vast empire.[3]

Similarly, the Roman Empire's meticulous attention to its postal service, *cursus publicus*, was one of its greatest strengths: It built a vast network of runners that allowed for lightly garrisoned outposts to quickly call in additional troops if threatened. The power of its state was directly linked to its ability to share information quickly. Fast news was the backbone of the government's rule.[4]

Yet information had a hard speed limit—the humans and animals who carried it. As late as 1516, one of the fastest news routes in Europe was a postal service linking Antwerp to Rome. Using this blazingly fast service, living in Rome, you could learn about the breaking news in Belgium, nine hundred miles away, a mere eleven days after it happened.[5]

These postal services were the first news networks. The successive speed of these networks determined who could take power and keep it.[6] Because of this, investments in information networks were huge, for commerce and for governments alike. If you ever find yourself wondering why old central post offices tend to be grandiose, stately buildings with columns and atriums, this is why. The post office was a reflection of state power and a way to show off.

Humanity's information networks have improved in two key ways over the years: speed and accuracy. We can look at the history of news within these two ever-increasing but contradictory variables. They are contradictory because the faster the news spreads, the more likely it is to be incorrect. This is an inherent tension in all news: Despite its inclination toward inaccuracy, we want news to be fast and immediate. We crave the new, and yet we also want it to be truthful.

Our current obsession with virality is the same thing, with the same pitfall. Viral news tends to be inaccurate, but we still want it. It's the culmination of an age-old desire for what is novel.

But news does something else for us. When we look to the news, we are spending time with society. We are trying to understand and care about the wider world. It's space for us to figure out what's important, not just to us, but to our whole species. When we consume the news, it is a moment of dedicating our attention to humanity.

Natural Deadly Rumors

It's tempting to think that viral misinformation is a modern invention of social media and malicious actors. Yet fake news is as old as news itself. Throughout history, falsehoods have been shared widely as facts and stood uncorrected for months or years, regularly becoming accepted truth.

Many of these stories were largely consequence-free, such as the widely believed report in 1569 of a Leicestershire woman who was "confirmed" to have given birth to a cat. Others led to tragedy and horror, such as viral rumors that the Black Plague was caused by Jews poisoning wells, which led to executions and violent pogroms throughout Europe.[7]

Before the establishment of verified news systems, most news had a moral shape. Until the Enlightenment, collective sensemaking was done largely through religious means. The birth of conjoined twins, for example, would be regarded as a grotesque sign of the parents' moral failings and sinful acts.[8]

When the steeple of the largest church in England, St. Paul's Cathedral, was struck by lightning in the mid-1500s, this wasn't seen as a natural disaster but instead as a divine judgment upon the newly restored Protestant church. A Catholic pamphleteer wrote this up as proof of the moral sins and God's due wrath for the abolition of Mass. A Protestant bishop's response instead called this a divine sign that Protestant reforms weren't happening fast *enough*.[9]

Our ancestors looked for evidence of a higher power among life's uncertainties. The root of this was a long-established and consequential attempt to understand the world through our moral impulses.

What we might today consider moral or religious misinformation was simply the news, and it was the best you could get. This type of superstitious and sensational sensemaking was standard fare. Explanations injected with moral significance reigned as truth.

Those responsible for sharing current events regularly tinted and embellished events. In Europe, a major source for news was the literate clergy, who often wrote of events progressing within the purview of God's plan as miracles unfolding for the virtuous, or as a punishment given to the unfaithful. When pestilence flared, as it regularly did in these centuries, it was reported as unfathomable and divine retribution for the sins of a city, or the specific, unknowable will of a divine being. Viral rumors regularly spread widely and led to enormous suffering. Lynchings and ethnic persecutions based on such unfounded rumors were common. This was in itself a function of moral sensemaking. Before the establishment of empiricism, cause and effect was steeped in moral judgment.

Institutional authorities often reached for threads of meaning where there were none. Though stories were reported, they were rarely what we would today call facts. An unexpected eclipse or a comet in the sky was often reported as evidence of an imminent calamity or political upheaval.

Facts and Friction

Regardless of the era, rumors and falsehoods spread via two basic steps: first discovery, then amplification of unverified knowledge.

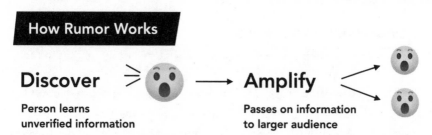

How Rumor Works

Discover ➢ 😮 ⟶ **Amplify** ＜ 😮 😮

Person learns | Passes on information
unverified information | to larger audience

The term *rumor* describes a type of information cascade, a proposition for belief which comes from unverified sources.[10]

Despite the wide proliferation of falsehoods, businesspeople, rulers, and politicians still required trustworthy knowledge and an accurate accounting of events. They would spend enormous sums on timely, credible news. For those under their employ—the earliest proto-journalists—sourcing truth was a constant struggle. For many centuries, the only reliable methods of transmitting news were patchwork systems of messages sent by couriers and sponsored by merchants, clergy, and nobility. This news was meticulously gathered, written, and verified.

Verification took many forms. For instance, before paper was widely available, the technology of parchment, the painstaking preparation of stretched animal hide, was used as a medium upon which to write messages. But it was expensive, and the grain of the hide itself caused the messages to be short and curt, not unlike the character constraints of early social media platforms. This often meant the messenger needed to carry the details and context behind the message itself. The parchment provided authority, and essentially said: "Trust the bearer of this note." Great care was taken to select the messenger—a verified human was necessary.[11]

Newsmen added what today we might call "friction" to the process of sharing knowledge by painstakingly validating messengers and stories through second- and thirdhand sources before they published them, lest they lose their reputation and sponsors.

Definition: Friction *noun.*

A limit or constraint placed on behavior or content to reduce its use or its spread.

This friction came to define the earliest news reporting. News that was both timely and accurate was incredibly expensive, requiring verified couriers and messengers. These were the first postal systems: Newsmen would write dispatches directly from the post and postmen would deliver news for a fee on the side. We can see this long-lasting holdover in newspapers that include *Post* in their titles.

Reach and Verification

The earliest American newspapers were far from reliable. Many of the first newspapers for general consumption competed for attention by aggressively peddling false, scandalous, or nakedly partisan stories, focusing on gruesome crime coverage in particular. But through the nineteenth century, some papers slowly matured and professionalized, building reputations for publishing factual narratives, engendering trust and status as "objective" news sources.[12]

During the lead-up to World War I, unchecked propaganda from all sides in the news reached a fever pitch, with every belligerent participating in a massive fight for public opinion. By the end of the war, it was clear that information warfare was a powerful weapon; it could raise armies, incite violent mobs, and destabilize nations.[13]

These learnings from the discipline of propaganda creation created a whole new set of problematic incentives for journalists after the war. The profession of public relations emerged, as professional marketers began using these new tools to sell products and manipulate public opinion at scale.

In response to this systematic manipulation of the truth, in the United States there was a concerted effort to professionalize the industry and introduce even more friction through the creation of fact-driven journalism schools. Some of these schools were funded by several of the worst early offenders—notably Joseph Pulitzer, who created the Pulitzer Prize for journalism and endowed the journalism school at Columbia University after spending much of his life as a purveyor of sensational yellow journalism.[14] Through fits and starts, this patchwork system of news gathering and distribution became the dominant way we empirically verified information before amplifying it. We learned to trust those that write in newspapers, largely because they fact-checked rumors. This process was ushered forward by the advent of the first mass-media communication networks: national newspapers and national radio. These slowly gave way to television, and between these three new platforms, a global media system took hold buoyed by the tenets of journalism.

This process of adding friction to the natural spread of information is a core part of the news industry today.

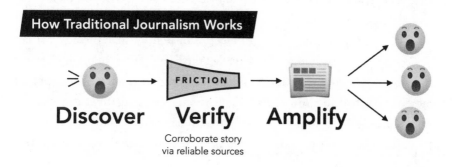

How Traditional Journalism Works

Discover   **Verify**   **Amplify**

FRICTION

Corroborate story
via reliable sources

With the emergence of radio and then television, news continued to evolve. And although these technologies allowed for unprecedented reach, they still relied on human gatekeepers. The public, a captive audience, was largely exposed to the same "objective" information.

The journalism of the American network era was fundamentally more powerful, one in which three major players, NBC, CBS, and ABC, controlled the majority of American attention. They, along with a handful of regional newspapers, set the news agenda for the whole country. Their sourcing mechanism—what might be called their editorial slant—was pulled from a narrow number of editors in smoky boardrooms who made decisions about what people should and shouldn't see.[15]

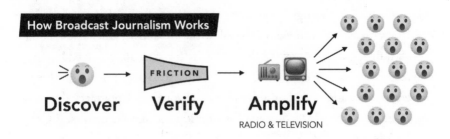

This system had major flaws. Reporting on powerful authorities, companies, and institutions was often uncritical, particularly if it might cause a conflict with the financial interests of the channel or newspaper. Government bureaucrats were often let off the hook in exchange for access to power. Having a single source of news had additional drawbacks; it was basically a monopoly, which allowed for fewer opinions that deviated outside of the mainstream. But facts were verified. And when politicians did things that were wrong, journalists still competed to ask questions and report on it—and scoop each other on the facts of the matter.[16]

This huge platform was so influential in politics that there was a law passed in 1927 in the United States called the Equal Time Rule, which stated that any candidate for political office who was given a prime-time spot for radio or TV had to allow their opponents equivalent airtime. Broadcasters were obliged to accommodate candidates to balance this power and at least partially offer cross-partisan opportunities, mostly to the two main parties, for editorial perspectives.[17]

Through this monopolistic media era, most professional reporters generally adhered to journalistic standards when it came to the proliferation of blatantly false viral rumors. This kept the fog of misinformation largely to a minimum, and viral falsehoods didn't spread widely through the population.

As Jonathan Rauch describes in his book *The Constitution of Knowledge*, we can think of this system of journalism, built over a century, as resembling a giant funnel. The wide end takes all sorts of information: rumors, hearsay, juicy gossip, conjecture, and more. Through a process of curation and corroboration, and critical opinion, the tiny quantity that reaches the other side becomes knowledge of real events, or news.[18]

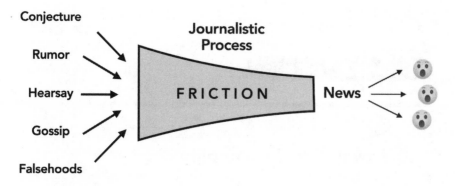

The epistemic funnel.

Every adult alive today has lived most of their lives within a modern incarnation of this information ecosystem, one that painstakingly, if imperfectly, balanced truths over falsehoods. We were all born into an epistemic environment that was built to privilege accurate information through the use of friction.

The Viral Era

In ten short years, the internet and social media blew this funnel to pieces. First the internet transformed publishing. In the late '90s, blogging platforms enabled anyone to publish whatever, whenever, without the critical eye of a journalistic colleague. Publishing was now a democratized, zero-cost endeavor.

When social networks emerged, distribution and reach were also transformed. Within a decade, hundreds of millions of people found themselves perpetually online in new, targetable, frictionless communities. Groups became digital gathering places for ordinary people, and not gatekeepers, to share information. The single-click Share button turned people into active participants in the distribution and amplification of information. News feeds pushed out bite-size posts to friends, and friends of friends. Curation algorithms used likes and favorites to decide what to showcase, and recommendation engines boosted engaging content even further.

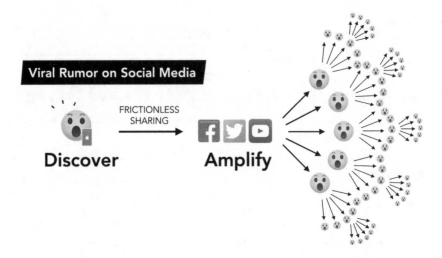

Viral rumors today will often obtain greater reach than traditional media broadcasts.

Reduced friction has enabled important new voices to be heard, but it has also led to the rapid spread of significantly impactful viral misinformation. When we entered the viral era, we increased the speed and spread of misinformation, without upgrading the system of verification that kept our knowledge stable.

The 2020 election, for example, saw far-fetched false narratives about stolen elections and CIA supercomputers going viral within as people just shared, and confirmed, the rumors that "felt" plausible. QAnon grew from a small online conspiracy to a decentralized online cult boasting millions of members, who energetically spread nonsense theories about corporations that the community alleged were involved in child trafficking. The COVID pandemic saw demonstrably, unequivocally false videos like *Plandemic*, which espoused numerous lies and conspiracies, reach an audience of millions before platforms decided to take it down.[19]

As a result of all this, we're living through a transitional period in history. Reduced friction has enabled important new voices to be heard but has also led to the rapid spread of fundamentally dangerous viral misinformation. We've upgraded rumors and downgraded journalistic friction that kept them in check.

Signal from the Noise

Our current location in time doesn't reveal much about the history of facts and falsehoods. If we really want to understand how we got to this current moment, we need to reach far back in time. As we zoom out, we can begin to see fundamental patterns emerging in how we share and process new information.

Each time a new media technology is introduced into our lives, we go through an enormous transition. Chaos and misinformation explodes, and people don't know what to think. Some people take advantage of this chaos to advance their own agendas. They accrue power and wealth. Eventually, they are held to account by people and institutions that check their influence by calling them out when they are wrong. We have come to call these people journalists.

But journalism didn't just emerge wholly formed. It took centuries to develop, with many unsteady efforts at building the institution of collective fact-finding.

The famous twentieth-century writer and political commentator Walter Lippmann had a theory about the natural progression of the institution of journalism. He believed that the press would pass through four phases and become more modern over time, marching toward becoming an institution that we can all trust. Within each of these phases, as time passed, the press would also become more objective and reality based.[20]

Lippmann wrote of four distinct steps, which I have broadly adapted into four stories from different eras.

The first phase is state monopoly, otherwise known as authoritarian control over all organs of the media, news, attention, and belief. This is a system in which the only narrative allowed is one that supports those in power. I illustrate this phase by exploring pre-reformation Europe, when the Catholic Church owned and controlled all narratives. This authoritarian narrative control collapsed spectacularly in European society due to a new technology called the printing press. The printing press detonated across the continent, and fundamentally changed the religious and moral identities of everyone it touched. We will travel back in time to understand what it was like to live in that era of profound turmoil.

The second phase was one in which political parties controlled the narrative. Activists and advocacy groups fight with political narratives and focus on what is most important to them. This was the state of the early press in America, one largely funded by partisans and activists with agendas. Here we'll explore the early outrage machine of the colonial American press, and how it shaped the world we live in today with the unique use of propaganda.

The third phase began in the 1830s when the press became supported by advertising and the public interest. This was driven by the commercial market of the public's appetite for news. We'll explore the paradox of how advertising incentives brought us sensationalism, and also created nonpartisan newspapers for the first time. This era gave birth to the news industry as we know it.

The fourth and final phase (which Lippmann himself thought he was ushering in) was the phase of professionalization. This is the world we were all born into—one in which the news became the responsibility of trained professionals alone, who were accountable for capturing the best version of reality and objective fact for the wider public. I explore this with two stories, one about the perils of early radio and television.

While Lippmann believed that journalism was entering its final stage of being a fact-based institution we could all trust while he was alive, I believe he may have been wrong. I believe we're entering a fifth phase, which I detail in my final anecdote. The decentralization of technologies that allow for the hypersegmentation of audiences into niche groups of partisans is now the world we live in—one of politically motivated targeting, which has created a fractured narrative. Social media has advanced this new phase but didn't create it. This most recent phase was ushered in by the deregulation of media and the advent of cable news. Many stories from the following chapters are based on three notable books: Andrew Pettegree's *The Invention of News,* Michael Schudson's *Discovering the News,* and Hazel Dicken-Garcia's *Journalistic Standards in Nineteenth-Century America.*

We'll also unpack how each new media technology has a dark valley when it's introduced. This valley is when new actors exploit a new tool at the expense of society writ large. The benefits and the harm can come in equal measure. The printing press, the telegraph, radio, television, and now the smartphone and social media all have a valley society must traverse in order to make them work for us all.

Here we go.

Chapter 17

The First Twitter Thread

Almost exactly five hundred years ago, a middle-aged friar in the backwater of northeastern Germany got mad. He wasn't a likely rebel. A stout Catholic monk who had spent his life diligently studying religious doctrine, he was well regarded among his peers in the Church. There was no reason to believe he would upend the most dominant institution in the history of Europe and kick off one of the most violent periods the continent had ever known.

But Martin Luther was nothing if not obstinate. He'd become a monk only after studying the law, and he was very particular about finding biblical justifications for the way the world was. In the early 1500s, Europe was still stuck in feudal poverty, with vast wealth held by the Church and royalty. Most people had no schooling and were likely to meet a single person in their lifetime who could read: their local priest. Because of this lack of education, they relied upon the Church for knowledge, moral authority, and access to meager social services. Most important for Luther, he believed that members of the clergy were the shepherds of peasants' souls to heaven—a responsibility he took very seriously.

The Church had recently begun a campaign to raise money by selling indulgences—little slips of paper that absolved the purchaser of their sins, or the sins of their loved ones, in exchange for a sizable amount of money. Sort of like hall passes to heaven. A refrain among churchmen selling these bits of paper went:

As soon as the gold in the casket rings
The rescued soul to heaven springs![1]

Luther found this outrageous. These indulgences cost small fortunes, and parishioners were being told the suffering souls of their dead relatives were at stake. Many who could barely afford them were spending their meager savings on what Luther suspected were worthless scraps of paper. In response to this and what he saw as other ethical lapses of the clergy, he wrote down ninety-five points of disagreement—*Ninety-Five Theses*—and sent it off to the local archbishop. Half a moral tirade and half an attempt at scholarly debate, this tiny tract of writing would come to focus the ire of the whole military, political, and religious apparatus of the time against him.[2]

Luther's *Theses* today distinctly resemble a modern Twitter thread.

- To say that the cross emblazoned with the papal coat of arms, and set up by the indulgence preachers, is equal in worth to the cross of Christ is blasphemy.
- The bishops, curates, and theologians who permit such talk to be spread among the people will have to answer for this.
- Those priests act ignorantly and wickedly who, in the case of the dying, reserve canonical penalties for purgatory.
- Etc.

Each is a point made in under three hundred characters. Each makes a critical statement. Each advances a moral judgment about how an institution is broken. And just like a controversial Twitter thread, it pissed a lot of people off.

In this era, this type of pointed criticism might have easily resulted in a prison sentence and a terrible death. Less than a century before Luther drafted his *Theses*, a very similar movement to reform the Church had ended in failure. At the turn of the fifteenth century, an English priest named John Wycliffe had criticized the clergy's rules and questioned the authority of the Pope with his own set of theses. His followers were gruesomely executed for advancing his ideas. After his death, Wycliffe was shamed and excommunicated, while his body was dug up from holy ground and burned.[3]

The threat of punishment like this ensured such blasphemy was rare. If people did fall out of line publicly with the Church, it mostly resulted in swift condemnation and incarceration, after which the heretic would recant or be burned alive. The Church was *the* establishment. It controlled the media and the narrative, by punishment of death. It was that powerful.

But Luther was born at a special moment. He'd crystallized his views in a time for them to just overlap with the rapid expansion of a new technology platform, one that likely saved his life. It would also change the direction of Western civilization.

In the years before Luther wrote his *Theses*, a skilled goldsmith named Johannes Gutenberg had perfected a novel process of mixing lead and tin to create a durable metal for movable type—a set of reusable and interchangeable letters. This type was placed into a specially designed mechanical matrix where it could be easily set and replaced by hand and fitted into a screw press primed with oil-based ink. Together these innovations constituted a new type of publishing machine. Instead of printing forty pages per day by hand, a team of two working a printing press could now produce upward of 3,600 pages in a single workday—nearly a one-hundred-times increase in output. These modern printing machines had recently found their way to Luther's town and been used to reproduce the very same indulgences that Luther despised.[4]

But this novel invention did not have an ideology. The recently created print shops were competitive and commercially driven. They wished to print things that would sell but had little way to determine what was interesting to their customers. What they found was that printing scandalous things was a surefire way to make money.

At first, Luther expected only a handful of people to read his scholarly criticism. To him this was an internal Church matter. Instead, his *Ninety-Five Theses* were reprinted by the hundreds, then the thousands. An urban legend soon spread alongside his writing, painting a picture of a man on a mission: an incensed and righteous leader dramatically nailing his personal outrage to the doors of the church in Wittenberg.

This story and his *Theses* were read aloud in town squares, spoken about in taverns, and discussed after Mass. Quickly banned in some towns, his writings would be printed elsewhere with a scandalous title as a guaranteed sale for printers ("Banned in Rome!"). Within a year, Luther had become Germany's most well-known author. Within seven years, millions of copies of his works had been printed across the whole of Europe—demonstrating a core principle for the first time: that media loves controversy.

By the time the Catholic Church recognized the gravity of what was happening, it was too late. The publicity surrounding Luther's writing was too enormous to ignore. He was too famous to quietly jail or execute. When the Holy Roman Emperor finally declared Luther's excommunication for heresy, he was protected by his fame. A German prince, who recognized Luther's significance, orchestrated a phony kidnapping by masked highway robbers, taking him to a castle for safekeeping.

While in hiding, Luther poured himself into his work, incensed by the Church's failures in accepting his criticism and emboldened by his viral fame. He had many more scathing pamphlets printed. Through a lengthy undertaking, he translated the Bible from Latin into the common tongue of German. He thought this would provide the faithful with an original, objective, clean, and pure interpretation of the holy word they could easily understand.

Before this moment, only clergy and a select few wealthy elites had ever been able to read the book upon which the authority of the Church itself rested. Luther's fresh translation, printed and distributed widely, was an instant sensation, opening up an opportunity for laypeople to interpret the "word of God" for themselves as he'd hoped. The printing press had democratized the Bible.

To Luther's surprise and dismay, this didn't lead to a single objective reading of the text. Neither Luther nor the Catholic gatekeepers of doctrine could do much to stop a tidal wave of new interpretations. What was once strict institutional truth suddenly became one opinion among many. New prophets were now able to speak of *their* truths, citing specific passages in the Holy Book. Literacy exploded, along with an array of public questions about the nature of the established order. Education and opinions blossomed.

But injustices that had been hidden in plain view also became visible. Previously accepted norms became moral outrages. From within this cascade of interpretations, some self-taught preachers began speaking publicly about a novel concept: God-given human rights. They could point to passages in the Bible and come to their *own* conclusions. Since the farmers and peasants of the time were largely bound to feudal lands in a system that resembled indentured servitude, they began demanding modest changes to their working conditions, citing the Holy Book as evidence.

This dissent became a protest, and this protest became a revolt. In 1524, more than three hundred thousand German peasants took up arms demanding better conditions for the working poor. Taking a cue from Luther's *Theses*, a large group of them established their own list of demands, known as the *Twelve Articles*. These were basic requests: Peasants should be able to fire their local pastor if he misbehaves. Peasants should be allowed to cut some wood from the forest. Orphans and widows shouldn't be taxed. Each resembled a modest request for new rights.[5]

But the feudal princes did not like these demands. Sensing a loss of power and a dangerous new precedent, they rallied their much-better-equipped soldiers, and instead of giving concessions, were openly antagonistic to their demands.

Local printers wrote about these events, but they reported upon the peasants' demands as unreasonable and their conduct as barbaric. False stories circulated widely about violent peasant rampages. After initially preaching tolerance, Luther was left infuriated by these largely false reports. He became incensed by what he felt were widespread and flawed interpretations of the

Bible, and the misguided new revolution. He sided with the ruling princes, calling for the peasants to be cut down like "mad dogs." The princes mobilized their forces, and with Luther's blessing, as many as one hundred thousand peasants were ruthlessly slaughtered, and the revolution crushed.[6]

But however horrible this massacre was, it was just the beginning of the bloodshed brought about by this new printing machine. It's difficult to understate how much violence came in the wake of this explosion of information. In the centuries following Luther's printed works, upward of 10 million people were killed in religious wars that spanned generations and continents. These convulsions and bloody persecutions defined the whole of the seventeenth century to come.

Keep this era in mind as we explore a thought experiment. Imagine stepping through a magical portal back in time, finding yourself in London in the mid-1600s, more than a century after Luther's *Theses* went viral. It's one of the largest cities in the world, with a population of several hundred thousand people. It's smelly, loud, and filthy.

You notice two fundamental things: First, the world you enter is one of factions and sects, one in which religious distinctions are ubiquitous and rigid. Righteous moral fervor is everywhere, and identities are shockingly strong: Cromwell's Presbyterians (Puritans) are now powerless after a period of ascendancy; their long and bitter struggle now seems lost. More than ever, people are one faith and are not other faiths. The dividing lines between Christian religious beliefs—between whole peoples from the same country—seem etched in stone. Puritans, Seekers, Grindletonians, Fifth Monarchists, Muggletonians, early Quakers: these identifiers are on everyone's minds. Being a Catholic is officially illegal; practicing Catholicism can result in fines or imprisonment or death. Tolerance itself is literally a sin that is preached against. It is encouraged to convert the sinners around you to the "correct" faith. If they will not convert, it is acceptable to punish them violently. The state officially condones only one religion, Anglicanism, but many more are secretly practiced. Everyone has an opinion about what constitutes the "right faith" and holds it tightly. Fanatics hold the authority.[7]

The other thing you notice is that regardless of personal faith, everyone has bad facts. Misinformation is rife, and terrible rumors spread like wildfire, feeding simmering grievances and often exploding into violence. Pamphlets fly about, full of propaganda about sins, godliness, and the right and wrong way to understand the events of the moment in the context of the Bible. The state prosecutes people with the wrong prayer books, but people share banned works left and right. Everyone has an opinion about this censorship. Outlandish stories about others' faiths abound. Rumors suggesting a secret pact between Catholics and the devil—even about how the Pope is literally the Antichrist—are accepted as fact. At times accusations of witchcraft surface, resulting in sham trials and murders by angry mobs.

Standing in this strange period, you might try at length to explain why these beliefs are wrong. You might describe how in the future these norms would be appalling. In the future there will be peace between Christians! Whole countries of different faiths will be tolerant toward one another! You make every attempt at convincing those around you that they're mistaken, that their customs are inhumane. Most find your ideas preposterous and heretical.[8]

You are stuck deep in the dark valley of the printing press, and it will be decades until you, and society as a whole, emerges. Misinformation has created cascades of outrage, revolution, and confusion. New readings of the gospel have created myriad "truths" that no single person can resolve. Cunning leaders have taken advantage of this disorder, grabbing power and prestige amidst the chaos. In the midst of these threats, people hold on to new identities to stay sane and grounded.

This is was what the printing press wrought. Gutenberg's machine amplified Luther's criticism and fragmented the order of the Catholic Church into uncountable Protestant denominations. It brought about the Reformation along with spasms of violent revolution everywhere it took hold. These spasms subsided only after more than a hundred years of violence and many, many civil wars.

After decades of conflict, staring into the abyss, society began to adapt, and a new type of pluralism became a reluctantly accepted norm. After a century of villainization and bloodshed, a unique idea eventually began to emerge—something resembling religious tolerance.

But this disruption also helped to usher in a fundamentally different human relationship with knowledge.

With increased access to religious education came agency, ideas, and some of the earliest concepts of human rights. It was the beginning of a new era of understanding the world. This revolutionary printing machine, coupled with Luther's outrage, with all of its fundamental flaws, forced a moral reckoning on society. It bootstrapped the Renaissance and primed the European continent for the Enlightenment.

IN SUM

Martin Luther and his *Theses* clearly follow some familiar patterns:

- The printing press was agnostic. It was first used as a tool of those in power, but quickly found use among those who wanted to see radical change in society.
- Even from its earliest days, the media and controversy have been magnetic attractors. Society wants to share, and be a part of critical new information and ideas, especially if they are controversial.

The enormous fragmentation that happened in the wake of the printing press took power away from the central authority of the day and fundamentally reformatted the way society was organized. This led to more than a century of bloodshed, as people tried to learn how to manage this explosion of new information and ideas.

Chapter 18

American Outrage

The Press as a Political Tool

Let's jump forward in time a hundred years, and across the Atlantic Ocean to colonial North America. To understand just how powerful our media technology can be, we need only look at the origin of the United States. American identity was shaped by an early outrage machine: the colonial press, a news enterprise that resembled an activist organization.

As students of the Enlightenment, the founders of the United States were entirely dependent upon the printing press to disseminate their ideas to the colonies. They very much understood how news was a tool for the influence of public opinion. Many held positions as printers, writers, and newsmen. They were all prolific contributors to the colonial press. Benjamin Franklin, himself the creator of several newspapers, was obsessed with printing technology and its impact. Even at the end of his life, as the most famous living American statesman and inventor, he still signed his name with the title "Printer."[1]

In Franklin's time, quality original content was difficult to find, so he wrote much of his paper himself. Many of his published works were fanciful stories, bits of advice, and clever tales written by him under fake names like "Poor Richard" and "Silence Dogood."[2] Using an obscure pen name or pseudonym was normal; it was common for an anonymous opinion piece with stark political claims to be published in the local press, only to be shot down by another anonymous pamphlet a week later. The word "objectivity" was a feature that would not come to be associated with newspapers for another hundred years. Any clear distinction between opinion and news was difficult to discern, nor was it really expected by readers.

Yet American colonists still eagerly read them. The Puritan roots of the early settlers ensured that many children were given a strong early education in Bible study and writing. The exception was the southern territories, where literacy was less widespread and enslaved people were generally prohibited from learning to read. But in the northern colonies, literacy rates were nearly twice that of the rest of England.[3] As a result, newspapers and current events were regularly devoured by a hungry market of readers. This fostered a flourishing industry of local print shops.

The colonial press was empowered by a handful of newly established press freedoms. In the 1730s, a sensational trial had been brought against a popular printer named John Zenger in a libel suit filed by a corrupt royal appointee, the governor of New York. The corrupt governor lost the case, as well as his reputation. It set a precedent in the risk of bringing unpopular suits against newsmen, causing those in power to think twice about sedition and libel cases. More than anything, it convinced British authorities that colonial juries would not convict American journalists.[4]

This environment allowed for an emerging culture of biting satire, rich commentary, and strong opinion among American printers. They covered regional and European news soberly while also publishing pamphlets with a dizzying variety of ideas and perspectives. These press freedoms would come to shape the American Revolution and the future country itself.

In the mid-1700s, an enormous global war had just ended between England and France. The massive colonial conflict would become known as the Seven Years War, or the French and Indian War to American colonists. (This conflict was so far-reaching, and so vast in scope, that Winston Churchill would later refer to it as the real First World War.)[5] It was also mostly fought in the New World. England had succeeded in expelling the French from much of the Americas at great cost, and it was in huge debt from spending heavily to protect its colonial assets. With its coffers dry, the British parliament decided to levy a tax on the colonies to pay for this debt, thinking that the colonies—the direct beneficiaries of its war with France—should foot the bill. This law, the Stamp Act of 1765, was drawn up by Parliament, which would tax the sale of most paper products in order to fulfill the war debt.

For American printers, the law was a travesty. It required every newspaper and pamphlet to be printed on stamped paper purchased from London and paid for in physical gold or silver, not colonial paper currency. Seeing the makings of a costly headache and their profits threatened by unnecessary logistics, the printers simply didn't accept it. Collectively, they mounted an angry response. "Patriot" newsmen, as they called themselves, began printing critical opinion pieces with intense vocal opposition to the act in pamphlets across the thirteen colonies. As an industry, its income was most affected by the taxes levied by this law, and they almost universally came out swinging against it.

Printers engineered outrage through a variety of methods. The *Ames Almanack*—the most widely printed almanac in the Americas with a massive circulation of more than sixty thousand—began printing two prices to inflame consumer support: "Price Before the Stamp-Act takes Place, Half a Dollar per Dozen, and 6 Coppers single. After the Act takes Place, more than double that Price."[6]

Other printers drew upon more lofty ideals, borrowing from popular Enlightenment thinkers like John Locke, decrying the tax by referencing the "consent of the governed" and human rights.[7] Some of the most strident public language exhorting the concepts of inherent American freedoms and liberties can be found in these papers, including the earliest publicly printed rhetoric citing "British Tyranny" and opposing the concept of "taxation without representation."

Others were even more over the top. The *Pennsylvania Journal* loudly declared, on the eve of the Stamp Act's enforcement, that it would cease operations if it was forced to comply. It overdramatically compared the act to enslavement:

The first steps of oppression are detected...
A day, an hour, of virtuous liberty is worth a whole eternity of bondage...

Adorned with a custom skull and bones masthead, the paper ended the issue with a picture of a coffin, and an obituary declaring its own death. "The last Remains of The *Pennsylvania Journal,* Which departed this Life, the 31st of October, 1765. Of a Stamp in her Vitals, Aged 23 Years."[8]

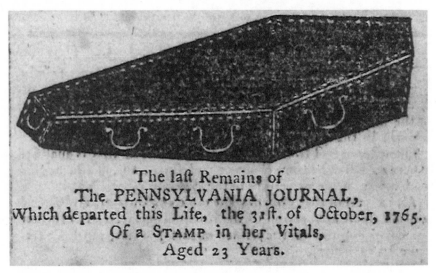

Source: The Tombstone Edition: Pennsylvania, The Last Remains of The Pennsylvania Journal, 1765.

As the historian Andrew Pettegree put it, "If the Stamp Act crisis of 1765 proved one thing, it was that the press is never more eloquent, self-righteous and clamorous in the defense of liberty than when its own economic interests are concerned."[9]

As these printers rallied against the Stamp Act, they circulated their outrage far and wide. It was difficult to be a colonist and *not* read or hear about the tyranny of the legislation. A new sentiment began to emerge, and opinion steadily shifted from indifference to grudging disapproval.

Many newsmen ventured even deeper into activism. Following the passage of the act in March 1765, a group began meeting at the *Boston Gazette's* office to prevent the act's implementation. They recruited a mob captain among the common people of Boston and began planning real violence.

On August 14, a mob was organized to taunt and threaten local officials tasked with implementing the law. One Stamp Act distributor, Andrew Oliver, was the inspiration for an effigy that was paraded through the streets, after which it was hung from the branches of what would later be called the "Liberty tree," beheaded, and burned. When Oliver didn't resign, the crowd turned to his office building and destroyed it before moving on to burn down his stable house. When word of this spread (also through the press), other riots began up and down the colonies, with similar mob strategies of burned effigies and sacked homes amid the bullying of local judges and business-people to ignore the use of officially stamped paper.[10]

This activist media campaign and subsequent outrage was so success-ful that within a year the act was repealed by the British parliament (along with a hasty law, meant to save face, that reaffirmed its right to tax American colonies in the future). Most colonists, satisfied that an injustice had been corrected, returned back to business as usual. The storm had passed.

But for others, it had just begun. This fury over the Stamp Act left in its wake a group of organized colonists with whom the language of inherent rights, oppression, and freedom had caught fire. Enamored with these ideals, and emboldened by their victory, they called themselves the Sons of Liberty. Propelled by the printers' outrage, the Stamp Act was the seed of American independence.[11]

This shows just how powerful our tools for sharing information influ-ence our perception of what is worthy of our collective outrage. Before we had a system of journalism, the colonial press operated much like an advo-cacy organization, not a truth-finding enterprise. It catalyzed through out-rage and changed the trajectory of the nation.

A Propaganda Machine in Boston

Just a few years later, the simmering anger was brought to a boil again. On the frigid evening of March 5, 1770, a tragedy unfolded in Boston.

The Sons of Liberty had done much to contribute to the mounting anger of the colonies, seeding unrest against British rule with tit-for-tat escalations and acts of subversion. When Parliament passed the Townshend Acts just a few years earlier—another taxation scheme for imported goods from Britain—the group organized more mass protests and riots up and down the colonies, culminating in a widespread boycott of British goods. This crippled trade with Britain in a number of areas, and within two years the British authorities had folded—again repealing a colonial tax. But as a punitive response, they sent a thousand new soldiers to be garrisoned in Boston to enforce tax and duty collections and quell further unrest. Perhaps unsurprisingly, it had the opposite effect.

Throughout the city, the mood was tense. A few days earlier, a series of brawls between local ropemakers and British soldiers had left one infantryman with a fractured skull. Among the locals there were rumors that an even larger conflict was imminent. Bostonians openly cursed the presence of the troops as an occupying force, and British regulars muttered of their desire to get payback for the brawl a few days before.

That night, the streets of Boston were covered with a foot of snow from a recent storm. A lone British soldier guarding the Boston Customs House sat at his post, shivering in his sentry box. He watched as a group of soldiers and colonists passed each other in the street. One of the colonists, a thirteen-year-old apprentice wigmaker, shouted an insult. The annoyed sentry, overhearing this, left his box and yelled at the young boy to stop his taunts, demanding he show respect. The boy, undeterred, turned and met the soldier, shouting a handful of insults at him directly. Infuriated, the soldier struck the boy with the butt of his musket and knocked him to the ground.

There was a flurry of shouts from observers as they ran to assist the boy, cursing at the soldier. A crowd soon gathered, yelling a string of profanities at the sentry. After hearing the commotion, more joined and the crowd grew, throwing snowballs and hurling insults. Retreating to the steps of the Customs House and loading his gun, the sentry yelled for help.

Learning of the unfolding crisis nearby, the commanding British captain ordered seven soldiers to the scene. They found the lone sentry pinned with his back to the door, with fifty angry colonists jeering at him and pelting him with chunks of ice and rocks. As they entered the fray, they formed a semi-circle with their bayonets as a barrier to the angry crowd. The ire of the mob now intensified, and they taunted the soldiers to fire their weapons. "If you fire, you must die!" they shouted. "Fire!"

As word spread at the fringes of the mob and through the town, a chain of viral confusion followed, becoming word of a "fire near the Customs House." Church bells rang—the signal rallying townsfolk to help put out a fire—and hundreds more colonists rushed to the scene, causing the mob to swell further in size. Facing a mass of humanity and vastly outnumbered, the captain shouted for the crowd to disperse and go home, insisting that no one would be hurt and that he had no intention of firing on the crowd.

As some members of the crowd began to tilt toward violence, they further pelted the soldiers with rocks, snowballs, ice, and clubs, leaving the soldiers nowhere to hide. An object thrown by the crowd struck the head of a soldier, who dropped his weapon, which then discharged. With the crowd still shouting "Fire!" in the confused seconds that followed, the soldiers fired their rifles in a disorganized volley into the mass of angry people. Five colonists were struck by bullets and killed. Screaming, much of the crowd fled into the night. Colonists and soldiers alike took stock of the horrific results. Smoldering anger had combusted into a tragedy.

In the tense days that followed, the flames spread up and down the colonies with an opinion war exploding in the colonial press.[12] By all accounts, the night was a disaster. But for a rowdy colony teetering on insurrection, the casting of blame was of enormous significance: Who was responsible?

Source: Paul Revere, The Bloody Massacre Perpetrated in King Street Boston on March 5th 1770 by a Party of the 29th Regt., 1770, Boston Gazette.

Members of the Sons of Liberty, seeing an opening, launched a pamphlet campaign against the soldiers, beginning in the Boston Gazette with the local engraver Paul Revere. He reproduced an etching of the event with a grisly—and deeply inaccurate image—showing soldiers in formation callously smiling and firing upon a crowd of colonial gentlemen.

This image, a study in propaganda, became a sensational and bloody cover image in sympathetic newspapers throughout the colonies.[13] The narrative of the murderous Redcoats launching an unprovoked, malicious attack on peaceful protesters was immediate front-page news. Empathy for the victims was asked, outrage toward the perpetrators was demanded.

But there was much at stake. Fearing a major reprisal and show of force from England in response, local authorities made an attempt to investigate and enact a fair and balanced trial. A young lawyer sympathetic to the patriot cause, future US president John Adams, decided to defend the British soldiers in court. He aimed to parse the truth from inflamed local anger. He hoped to turn down the temperature and prove that Massachusetts was a country of fair laws while avoiding a draconian response from the British.

A dramatic trial followed that instantly became a media sensation for the time. Adams was determined to find nuance in the midst of public outrage. He famously declared: "Facts are stubborn things; and whatever may be our wishes, our inclinations, or the dictates of our passion, they cannot alter the state of facts and evidence."[14] Through witness testimony and convincing oratory, Adams showed to the jury that the British captain had never given the order to fire. The soldiers had been cornered, trapped, confused, and beaten. The evidence laid bare by his defense, just two soldiers were convicted of manslaughter and branded with an "M" on their thumbs. But all were acquitted of murder.[15]

Yet despite Adams's reasoned defense and what was widely accepted—even among many Bostonians—as a fair trial, much of the propaganda had already spread. The narrative of the violent Redcoats and the helpless townsfolk had stuck. Empathy for the victims had demanded outrage. Context had been stripped from the event, and the Sons of Liberty had new martyrs for their cause. The incident came to be known as the Boston Massacre in the American press, and it was remembered as a bitter outrage worthy of retribution. Many were now swayed into joining the patriot cause. Colonists now had a rallying cry—a new national narrative of victimhood to organize around. The event decisively tilted public sentiment against England, setting the stage for outright revolution in the years just after.

Recalling moments of past outrages, we may be tempted to think that our modern sensibilities are immune to this type of reactive interpretation.

Before modern journalism, the outrages of the past were the result of small groups of people advancing a particular message and ideology—and forcing a choice on the large population. Our common national myths were shaped by such a group.

This sometimes had good effects. In this case, it hastened Britain's departure from the Americas and created the beginnings of a new national identity. It also created the space for a democratic system to form based on the tenets of liberty and free expression—core values of the press that have come to define our culture.

IN SUM

In this chapter, we see echoes of modern social media outrage in the media landscape that helped create the United States. Outrage played a role in shaping our national narratives:

- The colonial press was America's first outrage machine. The uprising against the Stamp Act, fueled by influencers in the media, laid the groundwork for increasing anti-British sentiment.
- A few years later, after a chaotic and deadly altercation in Boston, casting blame for the bloody incident became a media sensation, with anti-British propaganda stoking the flames through the colonial press. Despite John Adams's reasoned attempts to seek truth amidst the public outrage, a divisive take on the "Boston Massacre" had already gone viral.
- Outrage sowed the seeds of revolution, paving the way for a new system based on liberty and freedom of expression.

But an outrage-fueled system could not persist forever. In the next chapter, we'll see how newspapers evolved beyond the propaganda of the colonial period into the twentieth century. From here, we'll begin to see the rough shape of the first organs of journalistic sensemaking, subsidized by one very strange financier: ads.

Chapter 19

How Advertising Created Newspapers

What's the core issue with social media? Is it advertising? Many pundits and academics focus on the business model of advertising as being the primary driver of our problems with social media. Advertising forces our attention to be atomized, extracted, and sold. Advertising has been an enormous part of media distribution since the earliest days. It also was inseparable from the creation of the most significant fact-finding institution in history: journalism.

If any period resembled the modern cacophony of knowledge, and the explosions of opinion and questions about truth, reality, and objectivity in media, it was one specific era that began in the 1830s in the United States.

Modern news was basically invented at this time. Newspapers as we think of them today didn't exist before this point. Many of the epistemic issues we're facing at this very moment can be traced to a street in lower Manhattan, where a young man invented two things simultaneously: modern newspapers and modern advertising.

On September 3, 1833, a brash twenty-three-year-old entrepreneur named Benjamin Day printed the first edition of the *New York Sun*, and with it started a revolution. His new paper was dirt cheap and incurred losses immediately. But Day had a radical new idea: He believed he could turn a profit simply by selling ads.

The *Sun*'s slogan was "It shines for all" and cost only one penny, the equivalent of about a quarter today. Rather than focus on the less exciting parts of politics and higher society, as many of the six-cent competitor papers did, he instead focused on crime reporting and salacious gossip. Later editions would be illustrated and easy to carry, with vivid pictures of its top stories.[1]

This idea of selling people's daily attention to advertisers for a profit created a paradigm shift in news—one we're still living through today. This innovation fundamentally changed the nature of how humans acquire knowledge about the world. This launched a new industry of reliable attention capture, letting advertisers extract bits of our daily thoughts in exchange for information, while the news producers take the profits.

And profit they did. The first penny papers, as they were called, were tremendously successful, selling tenfold what six-cent papers did.[2]

Before the *Sun* opened shop, printers produced newspapers, but often at a loss. Printers made their living by selling printing services, or through political patronage. In fact, most early American newspapers (like the early revolutionary presses) were fundamentally partisan and politically oriented. Many were simply mouthpieces for political parties. They would toe a party line and share largely partisan news, and were funded through party affiliation, like the Democratic Republicans, Whigs, or Federalists. (Many papers still have these names as holdovers, such as the *Press Democrat* and the *Springfield Republican*.)[3]

These papers didn't have any pretense of political objectivity. In many parts of the country, the local postmaster was often the same person as the editor of the local newspaper. They ensured that copies of their papers were "franked" or delivered for free, but only their papers and their party's papers. (Imagine if today the Republican party's local newspaper was the only news you could get and a Republican postmaster also ran the post office, delivering most of your news. That would be the early news of postcolonial America.)

Early Americans expected that newspapers would be supported through this patronage from major political groups. Journalists of this era were usually "little more than secretaries dependent upon cliques of politicians, merchants, brokers, and office seekers," as the historian Isaac Clarke Pray said.[4] This rigid old media catered to elite crowds and did not think much of the general public's interest in news.

But the new pennies were unique. They found ways to split the interest of an otherwise indifferent audience by appealing to many diverse economic and social groups simultaneously: immigrants, artisans, and other ordinary citizens. News purchased through the penny press became the first affordable news to the general public.

By focusing on crime and sensational news, the penny papers carved out a new market for themselves that fell beyond the scope of the old newspapers. If previously, most newspaper articles discussed politics, reviewed theater, or published books, the pennies would focus on what was entertaining, or gruesome, or salacious. Benjamin Day was one of the first to go out and hire reporters to collect stories, sending them to the courthouse to gather details about local divorces, murders, suicides, and deaths. He was also the first to advertise his advertisement-driven paper, paying newsboys to hawk papers on street corners.

While the *Sun* was the first newspaper to fund its enterprise almost entirely through advertisements, it opened the door to a flood of competitors. The *New York Herald*, launched just two years later by Day's chief rival, James Gordon Bennett, became a fast challenger. It quickly shot to prominence by scandalously covering the shocking murder of a high-class downtown prostitute.[5]

The new penny presses were dedicated primarily to capturing attention. (As Bennett said, the goal of his paper "is not to instruct but to startle and amuse.")[6] As competition increased, so did the willingness to experiment. These competitors were often small shops. One person could be the editor, the ad agency, the reporter, and the printer of the entire publication, much like bloggers or influencers in modern times. Any enterprising soul could use a hand-crank printing press for a modest investment and, coupled with diligence and hard work, launch their own brand of news. For a bit more money, you could buy a new steam-powered flatbed cylinder press, which could increase your output to one thousand sheets per hour.

Amassing rapid audiences supported by advertising fundamentally changed the type of news that was reported. Sensationalism, crime reporting, gossip, and dramatic occurrences (many fictional) exploded across the news landscape in early America. Its value lay in the fact that it was interesting.

For the first time, newspapers began crudely competing for attention. These early papers found themselves with many of the same issues plaguing modern platforms. These printing outfits were one-stop shops for the viral news of the day. And not unlike the struggles of recent online media explosions (*HuffPost*, the *BuzzFeed*s and *Upworthy*s and other sensational news outlets that engineered clickbait in the early 2010s), they regularly stole whole stories and ideas from each other and repackaged them for consumption.

With the penny papers came an original race to the bottom both in content and advertising, as early papers saw a dramatic degradation of the quality of the companies and products sold on their pages. The early penny papers' advertising pages were chock-full of patent medicines and scammers trying to sell their wares coupled with exploitative claims, not unlike the ads we see today on YouTube and Facebook.

They were also righteous in defending their hands-off practice of selling this ad space (also much like contemporary social media companies). Then, as now, this type of laissez-faire morality was the result of shrewd business ethics.

Papers weren't quiet in criticizing each other's low advertising standards. The *New York Times* called the *New York Herald* the "organ of quack doctors."[7] (As historian Michael Schudson pointed out, this was the narcissism of small differences, as in the same issue, the *Times* published an ad for "The American Mental Alchemist," among other nonsense health aids and recommendations.)[8]

This resembles the type of morally agnostic advertising decisions that have been made at tech companies in the last decade. Twitter, YouTube, and Facebook struggle with similar influencer-peddled garbage medicines, from nutritional supplements to bogus vaccine substitutes.

When advertising became exclusively an economic choice rather than one that involved morals and standards, junk got through. It was a fundamental transition in how attention was captured and sold. Before this point, advertisements usually came in trade newspapers, which had titles of *Advertiser* and *Mercantile Press* in their titles but carefully guarded their reputations as "platforms" for quality products. The adverts *themselves* were the draw, and the trade papers allocated ad space with their morals intact. But for the first time, the penny papers made news itself the product, relying heavily on righteous drama to capture attention.

The penny papers made a case for moral authority as a way to keep eyeballs on their stories. They presented themselves as institutions, even if they were a handful of men in a shabby warehouse working by oil lamp. This remarkable pomp could be seen in their names, which suggested an air of noble benevolence to society: the *Tribune*, the name of an official responsible for protecting average citizens from arbitrary judgments in ancient Rome; the *Herald*, a royal messenger; the *Inquirer*, one that asks pointed questions. Others adopted the language of illumination: the *Sun*, the *Star*.

These tactics worked. The penny papers saw exponential growth, selling thousands of copies per day. After two years, between 1833 and 1835, papers with this new ad-based model outsold their rivals by nearly ten to one.[9]

Though current information was much more accessible, it was entertainment first. But in these cheap newspapers lay the first kernels of nonpartisanship; this was literally some of the first nonpolitical news. The salacious news subsidized real reporting for the first time. And through that, many of these newspapers eventually diversified and became a major conduit of public education. Society pages, helpful tips, cultural advice, and new ideas eventually constituted a new stream of learning and insight. For a working class with little formal schooling, these papers were also a new and valuable resource for personal improvement.

The penny press also created a new sort of shared social space in which private lives were no longer private. The historian Michael Schudson, in his book *Discovering the News,* speaks of how this played out in one particular article: Congressman William Sawyer's lunch habits were splashed across the pages of the *New York Tribune.* The "story" was about how he ate his greasy lunch of bread and sausage using his buck knife as a toothpick and wiping his hands on his coat sleeves. This was unheard-of in that era—writing a piece about a mundane, if unsightly, ritual in a public paper as a sign of poor character.[10]

But this created a new appetite for reporting on the personal lives of interesting humans and judging them harshly in the public square. A new sort of attention was paid to important people's private affairs, and public scrutiny pushed notable people to be more conscious of their image.

This public gossip began to build something akin to a common understanding of public civic life—what Hannah Arendt later referred to as *society*, "that curiously hybrid realm where private interests assume public significance."[11]

All the Lies Fit to Print

On August 25, 1835, the *Sun* began publishing a flagrant hoax. It was a fabrication in its entirety, falsifying the testimony of a real astronomer named Sir John Herschel. It claimed that he had discovered a species of bat-people living on the moon. Published in six separate issues over the course of several weeks, the discoveries on the moon were described in fantastic detail in the articles: winged humanoids with the appearance of bats. It said an immense new telescope was used to make these observations, also capturing unicorns, bison, and goats all living on the lunar surface.

This series was a sensation and a commercial success. These shocking falsehoods, printed in one of the most popular papers of the day, weren't revealed to be a hoax until weeks later, when a team of scientists from Yale came down from Boston to review the reports. The *Sun* never issued a retraction.[12]

It's no coincidence that the first ad-driven paper was also a paper that peddled egregious falsehoods. Many of the contemporary papers of the day, who also embraced the ad-driven model, began to delve further into the salacious, the ridiculous, and the outright false in order to sell copies. While the *Sun* had discovered a business model to support easy access to news, the news itself wasn't always accurate. But it wasn't just the medium of printed ink on paper. Images could lie, too.

Journalistic Photorealism

In the middle of the nineteenth century, a new technology began shaping the way humans thought about objective reality: photography. In 1853, a British photographer named Roger Fenton traveled to the Crimean Peninsula in Ukraine to capture images of the war raging between Russia and the United Kingdom and its allies. Over the course of his trip, he captured more than three hundred photographic plates that required long exposures. Though most of these photos were landscapes and portraits, one that depicted the aftermath of a fierce battle, with cannonballs littering the road, was more disturbing.

Upon his return to Great Britain, Fenton displayed the image, titled *The Valley of the Shadow of Death*, where it became an instant sensation. It came to be known as the first war photograph, heralded as an accurate depiction of battle, and captured the imagination of the public.[13]

In the decades to come, this photograph (and photography in general) would become a powerful symbol for journalism. The idea of the "lens" of journalism became both a literal description of the act of capturing events, and an effective metaphor for what journalists do: capture a faithful representation of reality. The observational nature of the newspaper could be likened to the way a camera shows things just as they are. Journalists began comparing themselves to cameras—accurate reproducers of truth. Since reality was "knowable" with this new technology of observation, a newspaper could do the same.

Years after his original exhibition, another plate was found that is likely the original photo Fenton took that day.

It shows the valley with no cannonballs on the road. A 2012 documentary exploring this discrepancy found that Fenton and his team very likely moved the cannonballs, spreading them evenly across the road to enhance the shot and make it more dramatic.[14]

The Valley of the Shadow of Death, Crimea, Ukraine, 1855. One of the the first war photographs. Researchers found that Fenton very likely took the empty road photo first, and then moved the cannonballs onto the road to create a more dramatic image. The dramatic version was what was published. Photos by Roger Fenton.

In this way, journalism was indeed just like photography. Not an objective frame of the actual event, but a theatrical retelling, a reconstruction of an event made to capture attention. A dramatized rendition of the original event, presented with a flourish to make it more interesting.

Fool Me Once, Shame on You. Fool Me Twice, Shame on Me.

While the *Sun* is an example of the salacious nature of the early news industry, its practices are consistent with the history of how media industries capture attention.

In 1833, Benjamin Day and his contemporaries used a new, cheap type of printing press to build an innovative new business model based on salacious and sometimes false news, and with it they captured the attention of less sophisticated readers. In the process they educated those readers, gave them new ideas and insights but also sometimes fooled them.

Paradoxically, the penny presses offered both more sensationalism *and* more unbiased information. They discovered that there was a market for objective and verified information, too.

Entrepreneurs use new technology to exploit our attention, but that exploitation has limits. The public has only so much appetite for unreliable stories and falsehoods. Papers expend the currency of their reputation every time they make a claim that turns out to be false.

Year over year, month by month, consumers decide to pick the papers that they know to be readable and trustworthy. Though the *Sun*'s readership might have expanded greatly after the Great Moon Hoax, its reputation as an accurate source of journalistic knowledge and a reliable newspaper took a hit.[15]

This is a coordination problem for the general public, as people try to figure out whom to trust. It's based on a set of four conditions:

1. The public doesn't know what's true and must rely upon news publishers to find the truth.
2. The public has options, and can select among many news publishers to find the truth.
3. News producers expend a bit of their brand's reputation every time they publish a story.
4. If a published story ends up being false, the public will lose trust in that publisher and will avoid it in the future.

This basic process created a sort of self-correcting mechanism for papers that printed falsehoods. Diminished reputations hurt sales and eventually put those publishers out of business.

In 1851, sixteen years after the moon hoax, another upstart newspaper began in lower Manhattan, in direct competition with the *Sun* and *Herald*, with its offices right next door.

The publication was founded by journalist and politician Henry Jarvis

Raymond, along with banker George Jones. Its first edition was also a penny paper. Addressing speculation about its purpose, the founders wrote:

> We shall be Conservative, in all cases where we think Conservatism essential to the public good;—and we shall be Radical in everything which may seem to us to require radical treatment and radical reform. We do not believe that everything in Society is either exactly right or exactly wrong;—what is good we desire to preserve and improve;—what is evil, to exterminate, or reform.[16]

This paper built itself upon a reputation of factually reporting the news, and that brand came to matter, even though it was supported by advertising with the exact same business model as the *Sun* and the *Herald*.

The *New York Daily Times* came along and attempted to tack upward, giving people more reputable, sophisticated, and verified news. Its journalistic integrity was tested by the public day after day, week after week, month after month, and year after year. In 1857, it dropped the word *Daily* from its title and became simply the *New York Times*.

It staked its reputation heavily on new journalistic standards and verifiable facts. The paper didn't get it right all of the time, but week over week, it built a steady reputation of being accurate.

By the 1890s, after going through a restructuring, and in consequence of the extreme sensationalism of the so-called pennies, the *Times* steadily became one of the most popular papers in the country.

Capturing much of the sentiment of that moment, the attorney Clarence Darrow said in 1893, "The world has grown tired of preachers and sermons; today it asks for facts. It has grown tired of fairies and angels, and asks for flesh and blood."[17]

The news is a competitive market. But it doesn't just compete for attention; it also competes for accuracy and reputation. A newspaper that peddles only falsehoods isn't one that will gain much value. Falsehoods reflect poorly on the publication and become a liability for the newspaper owner—and for anyone who takes that information seriously.

Through this competition, a norm of factual reporting was established.

Newspapers were not afraid to call out the scandalous, salacious, overzealous, and incorrect stories of their rivals.

On a daily basis, the massive quantity of content created required a certain type of competitiveness among news organizations, one that created a new type of professional ethic. Standards began to emerge on their own, simply by dint of an industry needing some form to their amorphous processes of reporting and publishing daily news. By the latter half of the nineteenth century, a natural disposition toward accuracy and verifiability became the standard. This intense need for content required protocols to be established for journalists.

By the 1890s, a predisposition toward hard facts began to emerge as a result of these daily pressures—both reputational and pragmatic—of trying to capture the best possible news. This gave rise to the first journalistic standards in newspapers—things we'd recognize today in modern news articles, like the "inverted pyramid" of story writing (putting the most important facts first, followed by supporting facts and details).

Several new advancements in printing presses actually began to change the tenor of the papers as well. In 1843, Richard M. Hoe invented the steam-powered rotary printing press, which was able to print millions of copies of a page in one day.[18] With the switch to rolled paper, mass production of printed works flourished, as these presses could run much faster due to this continuous feed. But they were extremely expensive and required a larger circulation to support them. The constraints of how far the news could travel—and how many papers needed to be sold to support such a press—actually influenced what was printed.

Partisanship, previously a staple of the newspapers, was carefully edited out of the straight news-reporting portions of the paper in order to appeal to a wider audience of mixed political persuasions. Newsmen wanted to sell more papers, and political bias tended to turn off their varied audience of news consumers. In this way, the decision to become nonpartisan was partially a competitive one, driven by new tech and economics as much as editorial influence.

In lower Manhattan, the *New York Sun*'s office was on the corner of Frankfort and Nassau Streets, across from City Hall Park. This street was the birthplace of the first major attention marketplace. The *New York Times, Tribune,* and *Herald* all sprung up around their first office, with the block eventually earning the name Newspaper Row.

The original incarnation of the *Sun* is no longer there. Long ago, the office was razed to make way for an on-ramp to the Brooklyn Bridge. But walking the streets, you can still sense the type of energy on the cramped streets of lower New York where this radical business idea was born. You might still get a sense of what it was like in that era, when newsboys would stand on street corners, hawking papers with extraordinary headlines. You might feel the static electricity of people shuffling past each other, hearing ideas, yearning for gossip and news, and for the first time being able to afford it.

In many ways, the internet today resembles this cramped, outrageous early news environment. The innovation—capturing our attention and selling it for a profit—created an industry that still informs how we see the world today.

IN SUM

Back in chapter 4, we explored why many blame advertising for the demise of fact-based information in social media today. In this chapter, we learned that dynamic is not new—we've been here before. The rise of advertising-based business models for news in the nineteenth century initially fueled a dark valley of misinformation, but then unexpectedly laid the groundwork for modern fact-based journalism.

- Early American newspapers were partisan and controlled by political party loyalists, catering to an elite audience, but they were largely unprofitable. In contrast, the penny press trafficked in crime reporting and society gossip to attract the masses, growing an audience for its advertisers. Blatant hoaxes and sensationalism eroded credibility over time, and readers began to seek reliability in their news.
- For newspaper owners, a predisposition toward facts became a pragmatic consideration to preserve reputations—and profits. Partisanship was separated from reporting to make newspapers more appealing

to a wider regional audience. By the end of the nineteenth century, newspapers like the *New York Times* had built a reputation on factual reporting and grown in popularity.

- A critical takeaway is this: ***clear reputational risk for news producers, coupled with strong consumer choice,*** helped print news emerge from the dark valley of hoaxes and sensationalism, into the system of sensemaking we know as journalism.

This journey through the dark valley isn't limited to print. Up next, we'll learn about the dark valley following the advent of radio technology and the government regulation that emerged to address it.

Chapter 20

The Dark Valley of Radio

If you switched on the radio while sitting in your living room any given Sunday evening in America in the 1930s, there's a strong chance you'd be greeted by the lilting voice of a Catholic priest from Detroit, explaining economics to you.

There's a good chance that in that hourlong broadcast you would hear him, with his gentle manner, speak of social justice as being the most important issue in the world. You would hear a sermon about the worker's universal right to organize. You'd also hear him rail against the Ku Klux Klan, which reportedly burned a cross on the lawn of his church. You also might hear him speak about focusing the country's attention on American issues at home, and quoting George Washington about avoiding conflicts overseas.

If you listened just a bit longer, you might also hear him speak about a subversive plot, brought about by a powerful cabal of Jews who controlled American banks, to make America a communist state. You might also hear him openly praise the new policies of the German chancellor Adolf Hitler, and denigrate the sitting president of the United States, Franklin Delano Roosevelt. He might also warn you of a deeper conspiracy to control the minds of the American public through the mainstream media.

You'd be listening to Father Charles Coughlin, one of the most popular radio broadcasters of his era, and one of the world's first radio celebrities. At the height of his broadcasts, his shows reached a full third of American homes. It was funded by "the radio league" a sort of proto–public radio station, founded by Coughlin and supported by tens of thousands of small donations of a few dollars at a time.[1]

Radio at that moment was the Wild West. Anyone with the money and knowledge could build an antenna and broadcast their programming as far as the signal would take it. Coughlin was an entrepreneur with an ear for celebrity, and he built the whole of his radio empire on hot-button political topics. He "platformed" himself, using the unique new medium of radio, and built an enormous audience on the back of historical grievances, emotional polemics, and political demagoguery. Using radio and his First Amendment rights, he created a media empire and one of the most popular shows in America.

Across the Atlantic, during this same period, another powerful orator, and the subject of Coughlin's admiration, had recently captured the minds of a nation using the same technology. Soon after seizing power and declaring himself Führer, Hitler and his chief propagandist Joseph Goebbels, had acquired all of Germany's radio interests, and they began a comprehensive campaign of making sure nearly everyone in Germany could listen to the Nazi party's broadcasts.

Historically expensive, radios were still out of the realm of affordability for average Germans. Recognizing the power of this medium to advance the Nazi cause, Goebbels worked with German engineers to build and subsidize Volksempfängers, or "people's receivers"—a radio that was affordable and could be mass-produced. It became tremendously popular and a cornerstone of their political control.[2]

The Nazis used their radio stations to transmit propaganda 24/7, domestically and internationally. They broadcast Hitler's speeches, as well as opera and dramatic shows. The historian Oliver Rathkolb called the use of radio a vital element of success in

Advertisement for Nazi Germany's Volksempfänger, or "people's receiver," 1933.

spreading Nazi doctrine in a way that simply "could not be ignored by the majority of the German population."[3]

Back in the United States, Coughlin's sermons inspired a huge political movement around the country, and fueled American sympathies toward the Third Reich while fostering deep hatred of Jews back at home. Coughlin's media empire had expanded beyond radio; he also published a hugely popular weekly magazine called *Social Justice*, which became a repository of his ideas and writing, even plagiarizing a speech by Joseph Goebbels under his own name.[4] Coughlin's innovative use of the radio made him one of the most important political figures of the era.

What happened to him? Coughlin's growing antagonism toward the Roosevelt administration and increasingly incendiary rhetoric caused alarm in Washington. After Kristallnacht, the night of violent pogroms against Jews throughout Germany, a number of stations canceled their syndication deals with him. Others forced him to send transcripts before they would play his broadcasts, reducing his reach and putting friction between him and his audience. In response, thousands of his supporters picketed in front of the radio stations that had banned his show.

The Roosevelt administration determined that despite the First Amendment protecting Coughlin's free speech rights, those rights did not necessarily include the medium of broadcast radio. In the government's view, the radio spectrum was part of the public commons and could therefore be regulated. Because of Coughlin, for the first time the US government required all broadcasters to seek operating permits for use of the radio spectrum. To comply with the Federal Communications Commission, the National Association of Broadcasters wrote a rule into their code of ethics specifically to force Coughlin off the air, and his permit was denied.[5] The association of radio broadcasters that syndicated his show, recognizing they had a responsibility to act as well, also stopped distributing them to their audiences.

After the Japanese bombed Pearl Harbor, pressure rose on him to stop his incendiary broadcasts entirely. Eventually, his capacity to operate was curtailed by America's entry into the war, after which his license to use the postal service to distribute his magazine was also revoked.

These were wartime measures to reduce hostile propaganda, but they show a clear willingness and necessity of the federal government to curtail activities that could be considered subversive. Freedom of speech, in this case, was not the same thing as freedom of reach.[6] By the middle of the war, public sentiment had turned decidedly against Coughlin, but he still pushed to broadcast his show. His superior at the Catholic Church, encouraged by the Roosevelt administration, gave him a choice: cease broadcasting or lose his pulpit. He chose to stop broadcasting.

Partially as a result of Coughlin's dramatic ascendancy and powerfully disruptive sway, the Federal Communications Commission began carefully regulating what could and could not be said—first on radio stations, and then television. These guardrails would stay in place for nearly three decades.

IN SUM

In this chapter, we examined how the unregulated early days of new radio technology yielded a dark valley of bigotry and divisive content including radio propagandists, from American priest Charles Coughlin to Adolf Hitler.

Our emergence from this dark valley was facilitated, in part, by intervention from the government. Unbridled access to disseminate hateful and violent rhetoric were curtailed by new frictions like airwave access permits and pre-syndication content reviews. The ability to go viral was not an inalienable right, and the government implemented wartime measures to control who had access to radio as a platform.

Radio regulation paved the way for regulation of other technologies. Television rose in popularity during a tumultuous time in the twentieth century. Next, we'll unpack why today *feels* so terrible, even compared with some of the worst decades in recent history.

Chapter 21

Television, Chaos, and the Collective

On a bright day in 1963, which had begun like any other in Dallas, a motor-cade traveling through a downtown park was struck by several bullets fired by a sniper. In what would become one of the most infamous moments in American history, President John F. Kennedy was killed in broad daylight while his wife and hundreds of onlookers watched in horror. The day ended with most Americans learning that the president of the United States was dead. The assassin, caught immediately afterward by police, was then assassinated just two days later in a spectacle that was captured on film and played by newsreels around the world.

Less than five years later, in April 1968, Martin Luther King, the country's most important civil rights leader and a brilliant strategist in nonviolent resistance, was shot dead at a Memphis motel in almost the exact same way—by a sniper with a high-powered rifle.

Just two months later, JFK's brother Robert, a presidential candidate, was shot and killed by a gunman at a victory rally in Los Angeles after winning the Democratic primaries for president in two major states.

In the months and years surrounding these events, the nation was roiled by enormous protests. Outrage against racial inequality and the worsening war in Vietnam brought hundreds of thousands of people to the streets to voice their anger about the country's direction.

Supported by an open draft, the Vietnam War was sending home close to a thousand dead Americans per month and killing thousands more Vietnamese. In 1970, during a protest in Ohio against the expansion of the war into Cambodia, National Guard troops opened fire on an unarmed group of protestors mid-day, killing four students between the ages of nineteen and twenty at Kent State.

During this time, a number of domestic revolutionary splinter groups launched bombing campaigns all across the United States. Police stations in almost every major city were bombed, as was the US Capitol itself. As Bryan Burrough writes in *Days of Rage: America's Radical Underground, the FBI, and the Forgotten Age of Revolutionary Violence,* the FBI tallied 2,500 bombings on American soil in a single eighteen-month period during 1971 and 1972—almost five bombings per day.[1]

In terms of economic progress, it was also one of the worst periods since the Great Depression. Growth rates dropped dramatically and inflation reached record levels while wages stagnated. An oil embargo imposed by members of the Organization of the Petroleum Exporting Countries sparked a global fuel crisis and led to soaring gas prices and frantic panic buying, with gas shortages all around the country. Unable to get gas, people waited in huge lines for hours at a time to fill up their tanks.

In 1974, it was revealed that newly reelected US president Richard Nixon had committed egregious abuses during his first administration as president. Forty-eight people, many of them top Nixon administration officials, were tried and convicted of crimes in a wide range of clandestine and illegal activities. They had bugged the offices of political opponents; ordered investigations of opposition figures; and used the FBI, CIA, and IRS as political weapons. In 1974, Nixon resigned in an attempt to avoid an impeachment trial he was all but guaranteed to lose.

The next year, in September 1975, President Gerald Ford survived *two* assassination attempts on his life, made by different people, in one *seventeen-day period.*

Almost all of these profoundly chaotic events were announced to the American public by a single eminently popular announcer on the television: *CBS Evening News'* Walter Cronkite. If they didn't watch his broadcast, they watched it on one of two other networks, which covered these events in much the same way, in a somber and matter-of-fact tone. Television was a massive and monolithic centralizing force in American culture, forcing everyone to deal with this trauma together.

One can take almost any period in American history and find spectacles of violence and disruption. But the decade and a half between 1963 and 1978 was one of the most tumultuous, violent, and disruptive periods in American history. If we look back at the abundance of crises and compare it to today, that era was a uniquely chaotic and disruptive time. When I meet people who were alive during that time, I often try to ask them what it was like to live during that intensely volatile period. What did it feel like? How does it compare to today?

Trust in government and perceptions of government fairness

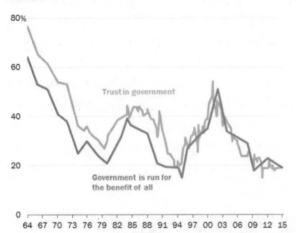

Survey conducted Aug. 27-Oct. 4, 2015. Q15. Trust trend sources: Pew Research Center, National Election Studies, Gallup, ABC/Washington Post, CBS/New York Times, and CNN Polls. Trend line represents a three-survey moving average.
Government fairness trend sources: National Election Studies and CBS/New York Times polls. Annual means calculated for years with more than one poll.

PEW RESEARCH CENTER

They usually say something like, "It was a pretty wild time to be alive, but today *feels* worse."

Looking back at this period, regarding the direction of the country, I try to calibrate the problems I see today and contrast them to what happened during those two decades of my parents' generation.

According to polling analysis by Pew taken over the last five decades, 1975 was a rough time for trust. Studies measuring public trust in government showed a decline, as did the belief in the concept that "government is run for the benefit of all." But public trust in the government still sat above where it sits today.[2]

Trust Falls

If the last decade has felt more traumatic, more toxic, and more outrageous than this disruptive era of American history, what changed? (Note: I write this referencing the years *before* the COVID pandemic hit.) How could we have ended up with such a profound difference in how we collectively feel right now?

The most obvious answer is that the way our media operated was different. People read major regional newspapers and watched the evening news through a tiny number of outlets. The issues of the day were shared, and the media wasn't structurally geared to stoke outrages.

In a 2009 study of different eras of newspaper columns, the authors of *The Outrage Industry*, Jeffrey M. Berry and Sarah Sobieraj, analyzed prominent newspaper columns in three different ten-week periods for 1955, 1975, and 2009. These eras were chosen assuming there was a precedent for outrage to be expressed in 1975—the height of the civil rights protests and national dismay about Watergate. Using 1955 as a control, what they found was that both 1955 and 1975 had profoundly low levels of outrage expressed in these newspaper columns compared to 2009.[3] Even in the midst of one of the worst political moments in our nation's history, people weren't exposed to as much opinionated criticism and outrages in the press. This accounts for some of the perceived difference in sentiment.

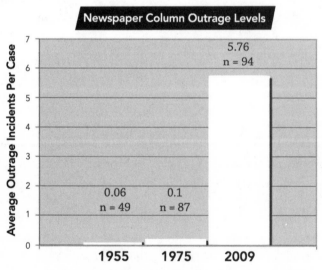

Newspaper Columnists' Use of Outrage Over Time

Source: Adapted from Sobieraj and Berry 2011

So what happened in that period between that caused such an enormous increase in outrages to be shared in the news? What changed the tenor?

Mad as Hell

In 1976, one of the strangest, most prescient films of the decade was released. The movie *Network* told the story of a faded network news anchor named Howard Beale—a relic of the glory days of the evening news—on the cusp of his firing from the network that had employed him for decades. The anchor begins to have a mental break and, on live TV, he announces that he's going to kill himself. The newscast ends with an enormous spike in viewership, and rather than firing him, the ratings-hungry network begins to "develop" his show as an entertainment program.

As his depressive state turns into a manic condition, his nightly broadcasts turn into rants about the injustices he sees in the nation, calling out the "bullshit" he sees everywhere. One night, clearly in the midst of a mental breakdown, he launches into a moral diatribe about the corruption of the state of the world, the problems of society, and passionately shouts at his viewers, telling them to go to their windows and scream, "I'm mad as hell, and I'm not going to take it anymore!" Cities across the country erupt as news viewers open their windows and scream into the night.

The TV network executives, tracking the ratings of this show and seeing an enormous profit opportunity, begin to orient the entire nightly news around his moral diatribes, incorporating a soothsayer, videos of real-life bank robberies, and sensationalized coverage in a live studio audience talk-show format. At the beginning of every episode, the audience screams Beale's tagline in unison: "We're mad as hell and we're not going to take it anymore!"[4]

The movie is a brilliantly written piece of fiction, with impressive, Academy Award–winning performances and plausible characters across the board. It also illustrates an amazing shift that was happening in broadcast news right at that moment, showcasing how ratings and emotional entertainment was subsuming the network news business for the first time.

Rage at the establishment and the system was an untapped new market for entertainment. It presaged the beginnings of trashy talk shows, punditry, and even reality television, all which would become staples of the American media diet in the decades to come.

Mad Together

Network beautifully illustrates the clear reason this period of chaos was somehow less destructive and polarizing for the country. People were still watching one of three networks: CBS, NBC, and ABC. As terrible, violent, and disruptive as this era was, Americans were still drawing from a common reservoir of content. If we were mad, we were mad together. The network news had become an ever-more-centralized business. Everyone was watching the same stuff.

A few laws passed earlier in the century and structural conditions made this centralization consistent for the average news consumer.

The *Equal Time Rule,* as we previously learned, was a law passed to address concerns that broadcast stations could easily manipulate the outcome of elections. This law was meant to reduce the influence of one-sided radio and television programs that presented just one political candidate in an election at the expense of others. This law is still in place today.

The other major law that dictated how content was served had more teeth: the *Fairness Doctrine*, which was passed in 1948 by the Federal Communications Commission. It required broadcasters to present controversial issues to the public in a manner that fairly reflected multiple viewpoints.[5] It didn't mean that individual TV or radio programs needed to offer a balanced view, but it meant that across a station's offerings, they needed to provide other programs that could balance that partisan programming with alternate perspectives.

The Fairness Doctrine was an attempt to reduce the monopolization of attention that could come from the enormous centralized power of media broadcasters. It was meant to encourage flexibility and access to more opinion. In practice the doctrine inadvertently reduced the incentive for most political programming to be aired on stations at all, because it required station owners to go out and find alternative programs and content that balanced the perspective. It actually reduced the frequency of political programming in general, because the task of providing multiple political perspectives was a headache.

The Partisan Piecemeal

But not everyone liked these rules limiting political conversation in the media. In the mid-'80s, many Republicans had generally become suspicious of a left-leaning bias in the press. After the Nixon administration's scandals, entirely brought to light and hammered home by mainstream newspapers, some conservative politicians suspected that the media was out to get them.

In 1987, with the new broadcasting technology of cable soon coming into mainstream use (and recognizing the new difficulty in enforcing it), Reagan's FCC repealed the Fairness Doctrine. This deregulation changed the media landscape almost immediately.[6]

Just one year after its repeal, a middle-aged talk radio host previously holding a regional time slot in Sacramento, California, launched a talk show with a new format. A shock jock and entrepreneur, he knew how to step up to the line, court controversy, and then deflect criticism away from himself. Without the Fairness Doctrine keeping his commentary in check, he could now launch a show that would court a new kind of controversy—*partisan* controversy—with no fear of losing reach in the syndication deals that make or break a national radio show.

His name was Rush Limbaugh. He said outrageous and partisan things because it only drove more traffic and interest in his show. Limbaugh skirted the line of the acceptable precisely *because* he knew it would improve his ratings. In this unique new regulatory environment, he was able to split the radio audience apart for the first time, peeling away a sizable chunk of listeners by appealing to their political biases.

He offered the show to radio stations across the country for free—as long as they contractually played his advertising segments. It was an immediate sensation, and by 1990, just two years later, he was regularly capturing five million listeners per show and reaping millions of dollars in ad revenue.[7]

Limbaugh was at the advancing edge of a new trend in punditry and journalism—appealing to moral outrages in order to sell advertising. His show, along with television and newspaper commentary, was going through a fundamental shift in approach—away from the "objective" frame of Walter Cronkite—and toward opinion journalism that made people angry.

Limbaugh didn't care as much about facts in his commentary as he did about the feelings his polemics generated. He had very little regard for journalistic standards. He invited speculation and conflict, stoked it, poked and prodded at it, and in the process invented a type of conservative punditry that was fact-agnostic. Limbaugh would regularly field partisan conspiracy theories, and can even be cited for popularizing the original false claims that Barack Obama wasn't born in America.

Because of his unconventional moral stance, humor, and general savvy, this wasn't a problem for his audience. Limbaugh's ardent supporters excused almost anything he did and deflected blame by claiming liberals were simply being hysterical. If that sounds familiar to observers of Donald Trump, it's because Trump was a close friend and careful student of how Limbaugh captured attention and presented himself. As the Harvard scholar Brian Rosenwald said to the *New York Times* upon Limbaugh's death:

"Without Rush Limbaugh, there is no way you get from the party of George H. W. Bush to Donald Trump. Over 32 years, he conditioned his audience as to what they wanted to hear and what they had an appetite for. And it thrilled them to hear someone who said what they might have thought, but felt uncomfortable saying... Trump applied that to politics."[8]

Limbaugh's show proved that there was an extremely profitable new market for opinion and commentary. In the following years, other media began experimenting with similarly biased shows with mass appeal. Most notably, Fox News.

When Fox News was launched in 1996, it quickly carved out a market for a conservative-leaning all-news channel, something that wasn't previously economically feasible because of the Fairness Doctrine and the restrictions of broadcast media. As cable television began expanding across America, it was possible to build niche audiences better than ever before with specialized shows and programming.

Fox News debuted with the tagline "Fair and Balanced," a differentiation that was a soft dig at the mainstream press and a nod to conservative claims of liberal media bias. With increasingly partisan opinion programming, the network grew in influence and size over the following two decades, and by 2016 was the most popular single news network in the country. It anchored an emerging right-wing media ecosystem, and, along with news blogs and conservative talk radio, gradually began sharing an alternate version of reality in its opinion lineup.

But Fox News still had reporters whose job it was to report facts. These were professional journalists who did good reporting much of the time. The news desk and the opinion desk were separate parts of the organization. While there continued to be a clear wall between the news desk and opinion coverage, the opinion programming steadily increased its time. The result of this was that after years of the winnowing-down of the news desk and the prioritization of extreme punditry, Fox News became more of a publicity wing than a news organization. As former Fox News chief political correspondent Carl Cameron said, "There's far more opinion than facts on Fox News."[9]

The 2017 book *Network Propaganda*, by Yochai Benkler and others at MIT, analyzed the way information—particularly political falsehoods—transit within the mainstream media and the right-wing media. In their study of how 4 million political stories traveled through social channels, they found something striking. There were asymmetrical differences between the left and the right.

The authors found that while both mainstream and right-wing media ecosystems had fringe conspiracy theories that regularly cropped up (from the Drudge Report to Breitbart to Fox News), including straight lies that often made their way to politicians' talking points, the mainstream media was much more effective at culling these falsehoods before they gained too much traction (such as a viral story about Trump having sex with children at Jeffrey Epstein's mansion, which was dismissed by mainstream outlets). Opportunistic conspiracies on the right, however, regularly made their way into the opinion reporting on conservative networks and continued to spread (such as the story of how democratic operative Seth Rich was killed by a Clinton-directed assassination). The right-wing media ecosystem would simply not fact-check falsehoods at the same rate as the left.[10]

What this suggests is that the powerful fragmentation of media that happened during the latter half of the twentieth century has been one of the primary forces shaping our breakdown of shared collective truth. The peeling away of an entirely independent right-wing media ecosystem and its continued willingness to flirt with a hugely profitable market for falsehoods has dramatically contributed to the problems we're facing.

This segmentation of new markets illustrates a pattern of increasing personalization of our information streams. This same focus on personal preference in our media is echoed all across social media as algorithms give us more of what we want and the opinions we disagree with the most. Our news begins to mirror our own moral foundations. As we struggle to determine facts from falsehoods, we become more willing to toe the line of our in-group at the expense of a shared set of facts.

IN SUM

The divisions we feel today on social media feel different than in the television era. Through the violence and turbulence of the mid-twentieth century, Americans were still largely forming reactions based on a common reservoir of content, enforced by media monopolies, airwave regulation, and laws like the Equal Time Rule and the Fairness Doctrine.

This changed with the repeal of the Fairness Doctrine and the emergence of cable television in the late '80s and early '90s. Entrepreneurs like Rush Limbaugh discovered a profitable new offering for niche audiences: extreme partisanship. This new content category appealed to moral outrage to build an audience and sell advertising, and the line between opinion and fact became increasingly blurred. With outrage driving profits, opinion journalism began to proliferate in emerging pockets of print, radio, and television media. This atomization of news consumption came to match consumers more closely with their moral preferences in news.

Throughout part III, we've looked back at the history of print, radio, and television to see the patterns of outrage that emerge when new technologies grow faster than truth-management systems. The path out of the dark valley can be long and bloody, as it was after the introduction of the printing press. The emergent industry of journalism, driven by market forces and advertising—alongside institutional regulation—helped us traverse dark valleys of radio and television.

In part II, we learned about the unique features that make our new dark valley exponentially more dangerous than dark valleys of the past. Making sense of the truth is an important requirement to exit any dark valley, but our collective sensemaking is under threat like never before. In part IV, we'll explore how our sensemaking works and why it's in crisis today, starting with an examination of the institution of journalism and how our current dark valley is causing it to fray.

In the next chapter, we'll unpack exactly how journalism succeeds (and how it fails spectacularly at times) in telling us the truth.

The Cogs in the Machine

Chapter 22

How We Learn the Truth

Everyone agrees journalists must tell the truth. Yet people are fuddled about what the truth means.

—Bill Kovach

In 1865, a skinny and headstrong young Austrian immigrant wandered around the docks in St. Louis looking for work. After recently serving in the American Civil War, he found himself intermittently homeless and nearly broke. He had spent the previous months doing odd work as a deckhand unloading cargo, filling boilers with coal, and generally picking up any job he could find, including shoveling the manure of unruly mules in a stable.

That day while roaming the docks, his fortunes looked to have changed. He was approached by a fast-talking man who offered work on a Louisiana sugar plantation. The man had a boat, and for five dollars per head (nearly $150 today) he said he could guarantee passage and well-paid work for him and a few dozen other men who were also in desperate need of jobs.

The young immigrant took the offer, giving the man nearly all his remaining cash, and boarded the smelly and cramped steamboat with his belongings. After floating about thirty miles downriver, the boat docked and the captain asked the men to disembark briefly so he could deal with a problem, which they duly did.

The man and the steamboat then churned off downriver without them. They waited for him to return, but as their patience turned to confusion then to anger, it dawned on them that the boat wasn't coming back. They had been swindled—the job and the boat ride were a scam. Stewing with rage, the men were forced to trudge thirty miles by foot back to St. Louis on their own, worse off than before.

Upon his return, still filled with fury, the immigrant sat down and wrote this experience into a story explaining what had happened. He offered it to a local St. Louis newspaper, where to his surprise and joy, it was accepted. The account was soon published, as both a warning to others and a check on the fraudster, who wouldn't be able run the scam again in St. Louis. The young man was paid for transcribing this outrage, the newspaper's readers were informed, and society was better off.[1]

This man's name was Joseph Pulitzer. He would go on to become a reporter, a politician, a publisher, and a newspaper owner, and he would ultimately shape the institution of journalism itself. Over the following decades his papers would become a check on graft, government corruption, and visible injustices, building one of the largest (and most sensationalistic) news empires in American history.

At an atomic level, Pulitzer's first story is an example of journalism at its best—when it provides a basic check on corruption in society. An eye on power and abuse, funded by average people's interest and attention.

Journalists provide three distinct services to the general public. *Corroboration*, *Curation*, and *Critical Opinion*. Each is key to our sensemaking, and each has serious failure modes.

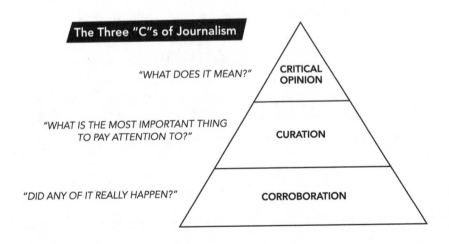

The Three "C"s of Journalism

"WHAT DOES IT MEAN?" — CRITICAL OPINION

"WHAT IS THE MOST IMPORTANT THING TO PAY ATTENTION TO?" — CURATION

"DID ANY OF IT REALLY HAPPEN?" — CORROBORATION

Collectively, these services are worth *billions* of dollars to society at large. They help us see hidden issues, stamp out graft, root out corruption, and generally call out those in positions of authority.

Yet Pulitzer also represents journalism at its worst when taken to its extreme. Later in life, his media empire would print highly sensational news stories and advance a kind of tabloid journalism that would terrify the public and stoke mass outrages. His public feuds with another media magnate—William Randolph Hearst—would come to be known as the scourge of "yellow journalism."[2] These headline-grabbing stories were hugely influential in politics and society, and led to many excesses, social anxieties, and moral panics.

Journalism is both of these things: It's a tool for understanding, and a tool for capturing attention. There are inherent conflicts within these two functions. If these sound like familiar tradeoffs, it's because social media struggles with them as well. Let's pause for a moment and compare the fundamental distinction between newspapers and social media companies. Weirdly, they might not be as different as we think.

The Handoff

We might envision the development of the internet as a series of baton passes in a years-long relay race. Not unlike the moment in a relay race when a new runner, with a fresh set of skills and new legs, is handed the baton from the one before, old institutions touched the internet and passed the baton forward.

Librarians and scholars, using old documentation tools to manage analog books and research (remember the Dewey decimal system?), passed the baton forward to a powerful new indexing service: Google. While Google didn't replace libraries, it gave us a faster and far more efficient way to source knowledge. Similarly, when encyclopedias met the internet, they passed the baton forward to a volunteer knowledge network run by thousands of people managing a citation system: Wikipedia. While Wikipedia is certainly less accurate than a published encyclopedia, it's thousands of times the size and available to anyone for free. With each of these handoffs, the embedded knowledge principles remained in the design, and we ended up better off.

When social media emerged, it seemed like something wholly new. There didn't seem to be a baton pass at all—it resembled a new invention entirely. We didn't realize it, but we had missed the baton. It was being passed from the institution of journalism forward to all of us—making us all journalists, but without our understanding anything at all about its principles. As a result, we've found ourselves running in a race we don't know the rules for.

Newspapers Are Social Networks

If you blur your eyes for a moment and examine the basic design of social media, and compare it to a newspaper, they are not so different. People—journalists or users—find, create, or share important and entertaining pieces of information they have found, and offer it to an audience. Someone—an editor or an algorithm—picks through those items and decides the rules by which content can be presented. The service then places space between this content, and then sells it to advertisers.

The primary difference between them comes down to two things:

- *The Serving Instructions*—What is allowed and what is not allowed to be placed in front of consumers.
- *The Scale*—How many people use the service, how often they are plugged into it, and how much the company knows about their audience.

For both newspapers and social media, as their scale increases, the serving instructions also increase in importance. But for a newspaper, the reputational, legal, and financial liabilities for serving dangerous content are still linked with their scale. You may note that this is not the case with social media.

This insight points us toward solutions. If we want to think about how to fix social media, we must look at the original platform of early newspapers. In Chapter 19, we learned that newspapers went from a cesspool of outrage, scandal, and confusion, and turned into an institution that sorts fact from falsehood. How exactly did that happen? If journalism has been around for more than a century, how does it still fail us today in determining what is objectively true? How might we use these lessons to build something better? What the hell is "objectivity," anyway?

As the algorithms we use today are becoming more and more influential in society, we need a basic set of principles for determining what they can and should do. Unpacking the failure modes of the older system may help us reverse engineer a set of fixes for the new one. Let's dive in, starting with our modern crisis of journalistic trust.

A Crisis of Trust

The viral era has brought so much new content into our lives that we've come to trust news outlets less and less. When we think of "news," we often think of what our news feeds give us. It doesn't matter if it comes from a pundit, a journalist, or your uncle.

This has created a fundamental shift in the way we make sense of the world: an increasing tendency to view any disconfirming knowledge with deep skepticism.

One of the reasons for this abiding sense of skepticism is that the news-consuming public doesn't really know what journalists actually do.

For over a century, it wasn't necessary to know. Until the internet, most people had few options when watching the news or reading a newspaper because of the monopoly of distribution channels. There was no need to explain journalism's mysterious ways, largely because no one had any choice in the matter. Journalists were glad to possess this power of authority, and happy not to really need to answer any questions about why they had it.

Whether it's because of lack of transparency, obliviousness, or lack of effort,

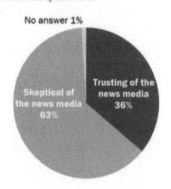

Most Americans think skepticism of the news media is good for society

% of U.S. adults who say that, ideally, it's better for society if the American public is ...

No answer 1%

Skeptical of the news media 63%

Trusting of the news media 36%

Source: Survey of U.S. adults conducted Feb. 18-March 2, 2020. "Americans See Skepticism of News Media as Healthy, Say Public Trust in the Institution can Improve"

PEW RESEARCH CENTER

journalists haven't taken the time to show their work. The public's lack of common knowledge of the *process* of journalism makes it easy to default to skepticism. Is there an agenda the press is concealing from us? Is there some greater deception afoot? This lack of general comprehension is the reason that the term "the media" today often appears as a pejorative.

Pick from any basket of common accusations voiced loudly in recent years: *The mainstream media is biased. Media executives have a left-wing agenda. The media is racist. The media is literally lying to you.* These criticisms about the media are cheap and plentiful.

We can sense that these institutions are teetering, perhaps past their expiration date, all while still projecting an air of authority, importance, and impartiality. It makes them extremely good targets.

But if journalism provides one thing that has yet to be replaced by social media, it's a serious attempt at providing a *verified reflection of current events.* Other than science and academia, few fields orient themselves around claims of capital-*T* "Truth" with such force and attentiveness.

Americans are often more negative than positive about the news media's role in society, standards

% of U.S. adults who say news organizations ...

Care about how good of a job they do	Are highly professional	Protect democracy	Stand up for America	Care about the people they report on
45%	33	30	28	23
37%	39	36	35	53
Don't care about how good of a job they do	Are not professional	Hurt democracy	Are too critical of America	Don't care about the people they report on
NEITHER APPLIES				
17%	27	33	37	23

Note: Respondents who did not give an answer not shown.
Source: Survey of U.S. adults conducted Feb. 18-March 2, 2020.
"Americans See Skepticism of News Media as Healthy, Say Public Trust in the Institution can Improve"

PEW RESEARCH CENTER

What the Hell Is Objectivity Anyway?

How can this claim to truth be substantiated? I wanted to know where this much-vaunted frame of objectivity came from in journalism. Is it a myth? An aspirational goal? A blatant facade? Trying to unpack it took me months of work to understand: extensive interviews with journalists and journalism professors, reading historical debates, textbooks on the history of the press, and many essays on journalism as well as dialogues with critics of objectivity.

I tried to find the organizational principles that define how the press works in practice. I discovered a shockingly misunderstood profession.

The reality is that journalism is a set of codes and norms, both implicit and explicit, established more than a century ago, baked into a body of creaky, once-formidable institutions driven by both profit and prestige. These institutions provide a largely misunderstood service to society: the service of *sensemaking*.

Today on social media, we're all journalists. We're all a part of the successes and failures of the enterprise of sensemaking. And it's impossible to look at our current media system as a separate thing from functioning democracies. Our opinions are always affected by the news, and our voting decisions reflect that knowledge. If you look at society like a big collective human organism, the news media is something like a central nervous system. It helps us respond to threats, share information, and figure out what needs to be fixed.

These institutions are under attack by politicians who now see journalists as a nuisance, and technologies that have empowered literally everyone with the power to report a story. Meanwhile, as journalism struggles to survive this multipronged onslaught, media institutions themselves are undermining their credibility in desperate ploys to stay alive and relevant.

This is a disaster. The result is a loss of trust in a free press, one of the most critical mechanisms we have for the correction of society's ills.

Let's dive into what journalists actually do. As sensemakers, journalists provide three distinct services to the general public: Corroboration, Curation, and Critical Opinion.

Journalistic Service #1: Corroboration—Did It Really Happen?

If your mother says she loves you, check it.

—journalists' saying

The first, and arguably most important thing news organizations provide is *corroboration*—the verification of events. This service provides a way to know the difference between rumors and reality. It underpins the entire enterprise. Without this, the entire system fails.

Corroboration serves as a tool to identify what happened in the recent past by sourcing additional details about an incident. Usually this happens through direct interviews and conversations with those present at the moment of the event.

The basic journalistic process of verification looks like this:

1. A claim is made.
2. That claim is given some degree of weight or credibility based on the reputation of the claimant.
3. The claim is confirmed (or debunked).
4. The claim is published and amplified.

Hearsay and spurious claims surround every event of significant importance, particularly political ones. Every occurrence, piece of news, and even scientific discovery involves a battery of different claims. Remember that in our "normal" information environment, falsehoods are more common than facts.

Journalism gives us this service of verification in order for us to understand reality. This is one of the reasons we pay for it. With major events, it's standard practice for journalists to find three separate independent sources to verify a claim. But for many other events, firsthand and secondhand sources will do.

Determining what actually happens in a contentious and chaotic event is hard, such as during periods of civil unrest or in the midst of a catastrophe or disaster. And since news is inherently new, the pressure to be first is sometimes at odds with the pressure to be right.

This balance between accuracy and urgency is core to the news-gathering process, and because of this newspapers don't always get it correct. Sometimes they will report on events too quickly, or source from the wrong place. Badly sourced stories don't usually make it to print, but if they do, *they become a major reputational liability for the reporter and the paper.*

As Bill Kovach and Tom Rosenstiel say in *The Elements of Journalism*:

> In the end, the discipline of verification is what separates journalism from entertainment, propaganda, fiction, or art...Journalism alone is focused first on getting what happened down right.

For Kovach and Rosenstiel, this forms the foundation of the discipline of verification, a set of maxims for journalists trying to figure out what is real: Never add anything that was not there. Never deceive the audience. Be transparent about your methods and motives. Rely on your own original reporting. And finally, exercise humility when reporting on what actually occurred.[3]

Journalistic Service #2: Curation—What Is Important?

The second service is *curation*. It answers the question: What should we pay attention to? Journalists and editors are answering this question for us, every day, when we read the news.

We must be engaged by the news in order to read it. In order to survive and thrive, news organizations must focus on their audience's interests. This makes them *select* certain stories over other stories. This is the process through which stories are determined to be "newsworthy."

Just as it doesn't make sense to regularly do front-page reporting on the victories of sporting teams living in other countries, news is effective only when it's relevant to us. Create content that readers *want* to read, and you'll maximize readership and profits.

But even between newspapers, there is a sizable difference in this selection of stories that might reach the front page. The order that they are presented in is its own form of editorial decision. Choosing what ends up as the front-page headline has its own logic and rules.

We can think about the selection of relevance like a simple set of instructions for a newspaper's editorial staff: Pick the stories that should go on the top half of the front page, so when the news is delivered, or the passerby on the street sees the paper, they will pick it up and read it.

If this sounds like rules for a modern algorithm, it is. Newspaper editors were the original engagement algorithms. These people regularly decided which stories were placed on the front page "above the fold" (i.e., above the crease in the paper). They were literally shaping our attention the same way a news feed puts certain content on top.

Media organizations maintain profits by providing consumers with the content they want to consume. Additionally, journalists are rewarded for writing articles and stories that capture attention.

Journalistic Service #3: Critical Opinion— Why Does it Matter?

Opinion, Analysis, Editorials, Punditry, Commentary, and Advocacy

Alexander Hamilton famously wrote, "Opinion, whether well or ill founded, is the governing principle of human affairs."[4] This *critical opinion* is one of the most important services that journalism provides. It's known by other names, including editorial writing, punditry, analysis, or advocacy. Its primary goal is to provide a strong perspective on events and occurrences, and to contextualize the news for an audience.

The two previously mentioned services (curation and corroboration) exist in the same room. That room is where "straight news" happens. Critical opinion happens in a different space entirely.

If you talk to any old-school journalist, they will tell you that "opinion" has no place in professional journalism. This may sound strange, but in news organizations there was once a clear boundary between the news desk and the editorial desk. If the owner of the paper wanted to tell you their opinion on a topic, they would write an editorial. If they wanted outside perspectives, they would pay someone to write an "opposite the editorial page" article, or Op-Ed (referring to its placement in the paper, directly facing the editorial section).

This is a bit difficult to reconcile with the views of most news consumers. It's hard for them to know the difference. If you talk to average people who consume news, they'll say that they turn to journalism specifically *for* opinions. Very often, when you ask them, "Who are journalists you pay attention to?" they will list a pundit, a commentator, or an opinion writer. In the mind of the public, opinion *is* journalism.

But in the minds of most old-school journalists, opinion makers should be in a very different part of the newsroom entirely.

The Services Journalism Provides

"Straight" ❶ **Corroboration** — *Verification of what actually happened*
News   ❷ **Curation** — *Selection of important events*

— — — — — — — — EDITORIAL WALL — — — — — — — — — —

Subjective   ❸ **Critical Opinion** — *Analysis, Editorials, Perspective*
Interpretation

This distinction should be clear, but even in practice it isn't. When we say, "journalism with a capital *J*," this generally means everything, from the weather to investigative reports, but with a careful carve-out for opinion.

People want these perspectives and seek them out. Strong perspectives are inherently interesting to news consumers. It is helpful sometimes for people to state forceful opinions, which can help catalyze useful conversations about the most pressing issues of the moment.

In the opinion section you'll find the "letters from the editor," an editorial board, and an official statement of the periodical's owners along with their perspective and analysis of events. It's also where you see opinion writers make bold, declarative statements about what should or shouldn't be happening in the world.

In practice, there is a very distinct dividing line between that part of journalism and straight news. Straight news is objective; opinion is not. This "wall" between opinion and straight news is one of the core tenants of journalistic objectivity.

A study of American news consumers by Pew Research showed that most struggle to tell the difference between straight news and opinion.[5] And, indeed on most news websites, there's barely any differentiation in the design or placement of editorial content from straight news. Much of the time if you see a headline with a strong statement, you may also miss the tag of "analysis" or "opinion" beneath it. Even less so in our news feeds.

This dividing line is critical to understand because beyond it is where we lose all pretense of neutrality. It's here where bias is openly on display. It's here where opinions matter and are encouraged if they're provocative. This is also where advocacy finds a home.

When the line of opinion, journalism, and straight journalism becomes blurred, readers don't understand the difference between them. It's the goal of activists and politicians to advance their particular agenda through opinion articles and op-ed pieces. This is the one space where activism is acceptable.

This doesn't mean, however, that opinion can circumvent the norms of factual reporting and verification. Opinion journalism cannot report falsehoods or state outright lies. Editorials, analyses, and opinion pieces still need to be verifiable. They shouldn't include lies and factual inaccuracies. Reputations still matter.

How Journalism Fails to Reflect the Truth

By its own standards, journalism often fails to accurately represent the truth. Journalists don't like to admit it, even if everyone else does.

There is an inherent conflict between journalists' desires to adhere to the principle of objectivity, and the market forces that attract readers to the news. The news is, despite that claim to objectivity, still a market-driven enterprise.

Growing up, children learn that a story consists of five W's: *who, what, when, where,* and *why.* We'd assume that journalists have a similar agenda when they are trying to report a story.

According to Stanford economist James T. Hamilton, this assumption is categorically wrong. Instead, he suggests that there are five economic questions that explain what comes out of a reporter or producer. The five economic W's are:

Who cares about a particular piece of information?

What are they willing to pay for it, or what are others willing to pay for their attention?

Where can you reach these people through media outlets?

When is it profitable to provide that information?
Why is it profitable to provide that information?[6]

Hamilton suggests that news creators don't wake up every morning and openly exclaim, "I really need to maximize profits for my company today!" But they do want to keep their jobs. They will still be influenced by these basic questions as they operate in the attention marketplace.

Regardless of their desires, these questions determine their survival in the business of newsmaking. Newspapers want to remain in business, and reporters want to keep their jobs. They will adapt to market conditions and the forces at play in society to stay alive. Answering those other five W's will make that difference.

In this way, a newspaper or television show is like an organism trying to survive. In chapter 19, we learned that fact-based journalism evolved from this very need for survival, helping newspapers fortify their reputations and attract readers seeking a reliable source for information. From an evolutionary view, news adapts to the conditions of its environment, including the selective pressures placed upon it. It must be profitable, or it must be influential to satisfy its owners.

When laws or technologies change around it, this "news organism" needs to adapt. These laws and technologies have been favorable to newspapers over the last hundred years, giving us a fairly objective reading of the facts. But in recent times, traditional journalistic standards haven't been as necessary to the organisms' survival. In chapter 5, we saw examples of how reliance on social media for distribution has directly weakened the application of journalistic standards.

If a newspaper can no longer support itself, it is forced out of business. It dies. Profit and influence are needed to survive. The secondary effects and benefits need to be underwritten to operate at scale, by the public (as is the case with NPR and the BBC), or philanthropists (as is the case with Jeff Bezos, who underwrites the *Washington Post*, or Steve Jobs's widow, Laurene Powell Jobs, who underwrites *The Atlantic*).

The symbiotic benefit of these organisms, like plants producing oxygen for us to breathe, is accurate knowledge for the masses. The by-product of their existence is a common cause, a shared reality, and an engaged form of democratic participation.

But the evolution of this symbiotic news organism doesn't inherently trend directly toward objectivity. If the landscape changes dramatically, objectivity might be sacrificed for the sake of staying alive, as has happened in recent years with many newspapers and media outlets.

We should not mistake news media's fundamental driving goal: They are not democratic institutions on their own; they are market-driven creatures, trying to thrive in a competitive environment. If they are unable to capture attention, they will die.

These are a few ways the organism of journalism fails to reflect the truth.

Journalistic Failure #1: Selective Facts, or Failure of Proportionality

Stories capture our attention. Data doesn't. While selecting one anecdote from a pool of less interesting stories, journalists bend our attention toward some topics and away from others. This anecdotal selection can be very frustrating to news consumers when they recognize that the broader pool of information they have available may not be a direct representation of what's actually happening in the world.

The inherent problem with this is *proportionality*. It's rare that the stories that end up above the fold are actually the most proportional issues of the day. For example, if deaths were proportionally covered, the daily headline would be stories in which heart disease or cancer took center stage. Any breakthroughs in the fight against these diseases should, per our general risk and the risk to our loved ones, be top news.

But these types of deaths are normal, accepted, and expected. They're not as interesting because they're not part of a unique trend or phenomenon.

Table. Number of Deaths for Leading Causes of Death, US, 2015-2020[a]

	No. of deaths by year					
Cause of death	2015	2016	2017	2018	2019	2020
Total deaths	2 712 630	2 744 248	2 813 503	2 839 205	2 854 838	3 358 814
Heart disease	633 842	635 260	647 457	655 381	659 041	690 882
Cancer	595 930	598 038	599 108	599 274	599 601	598 932
COVID-19[b]						345 323
Unintentional injuries	146 571	161 374	169 936	167 127	173 040	192 176
Stroke	140 323	142 142	146 383	147 810	150 005	159 050
Chronic lower respiratory diseases	155 041	154 596	160 201	159 486	156 979	151 637

Source: JAMA[7]

With the great exception of the COVID pandemic, the things that are most likely to kill us are the ones we hear the least about in the news. This sort of selective reporting creates a signal-processing problem for the wider public as people become far more concerned about things that are unlikely to hurt them. It also leads to a sort of grim math in the attention economy, in which the value of an average life becomes tied to the novelty of their manner of death. Nemil Dalal refers to this as *selective facts.*[8]

For example: Imagine opening up the news one day and seeing a story about a series of recent kidnappings. These abductions are happening to people with your exact first name. They may also be happening in places near where you live. In fact, the hair color, eye color, and clothing style of every abductee resembles someone that looks exactly like you.

You may rightly believe this is a threat to your safety. After hearing about this kidnapper, you click a link, search for more, and begin scouring the internet for more information about the crimes. You'd want to learn all the available details and the status of the investigation. You'd desperately want to know more.

Behind the scenes, the human editors (or the algorithms responsible for your news feed) are selecting from among a huge body of *potential* stories. However, they know you'll likely respond to a kidnapping that appeals to your biases and catches your interest above another kidnapping victim whom you might otherwise ignore.

We relate to those who look like us, sound like us, dress like us, talk like us. We're much less likely to read a story about a kidnapping or a murder of a person of a different race, speaking a different language. This is not because that knowledge is inherently less important to the world, but because it captures less of our personal interest.

For people living in minority communities, this stings. The journalist Gwen Ifill called this trend by a particularly biting name—"missing white woman syndrome," which illustrates a critique of how the news prioritizes coverage of missing, primarily white females above other kidnappings, particularly those of people of color.[9]

By selecting specific anecdotes from the wider pool of available ones, this curation process is prone to its own set of powerful prejudices. It's one of these particular areas in which we can distinctly see the difference between the editorial slant of various news providers. Simply by going to the front page of the *New York Times* and then switching to the front page of Fox News's website, we observe that they don't cover or prioritize the same news.

Newspaper editors make these decisions every day, determining what's *important* vs. *irrelevant*. This protocol of interest leads to a type of homogeneity. In coverage of kidnappings, the decisions about what to cover are often ones that appeal to the newspaper's audience. These conditions tend to correspond with the race, the ideology, the religion, and the affinity of the audience. If an audience is majority white, the kidnapping of a young white woman is likely to appeal to the inherent biases of that news audience. If that audience has a disproportionate (or controversial) racial preference, then those preferences will be on display in their news-consuming habits, too.

Inherently, we are programmed to care more about our in-group. This manifests in many problematic ways, like xenophobia, racism, ethnocentrism. But in a more subtle fashion, we do this all the time. It's called *homophily*: the tendency for people to seek out or be attracted to those who are similar to themselves.

Any group of news consumers has its own biases. Despite what you may have heard about inherent objectivity, news organizations will oftentimes do their best to *appeal* to many of these biases in order to sell papers, not challenge them. This applies to political coverage as well. Avoiding coverage of a contentious or damaging political event, and prioritizing a different story instead, can obscure this coverage from view and save face for the politician.

Journalistic Failure #2: Seeking Conflict Where There Is None

Conflict is a news value, and it shapes where journalists look. Conflict of a certain variety is engaging to readers. Listening to a conversation with two people agreeing is far less interesting and educating than listening to two people who disagree. As the media scholar and journalist Logan Molyneaux told me, "Conflict is to journalists what sugar water is to insects. It is unavoidable—it's the very definition of news. When we see conflict, we're all attracted."[10]

Conflict is a powerful mental heuristic for us consumers of news. It's a great shortcut for understanding where people's opinions are. If two humans are arguing well about a topic, they're much more likely to be bringing their best defenses of their perspectives. In this way, conflict is valuable for discerning opinions and getting to the center of an issue quickly.

But it also can cause us to think there are controversies where there are none. Pitting a scientific expert on a topic against a political hack, for instance, will make it seem like there is still room for debate on an issue (see global warming).

Journalistic Failure #3: Groupthink

News is expensive.

For the most part, the news media has limited visibility into the world's events. They only have reporters stationed in certain places and have a limited number of sources they can select from. A news organization must choose where to send its reporters, a process that has a finite cost. For this reason, there's a narrow band of visible events available to newspapers.

The Associated Press was established to share the cost and burden of sharing information and news across the world. It was started by five different daily newspapers in New York City in May 1846 to share the costs of sending reporters to the Mexican-American War. Reuters had a similar origin.[11]

If one news organization reported on a hot story and ran it on their front page as a national news item, they could "share" the scoop. Other editors would have a fair chance to run the same story as their headline, lest they lose the initiative and look out of touch. It was also simply easier for another newspaper to select that same story and send a reporter to write up their own version of events in order to cater to their audience. It was also a cost-effective technique: By simply reporting on what your competitor's paper was reporting on, you could save on the costs and effort of finding the scoop in the first place.

But this sharing of narratives also manifested in systems of preference as elite editors at national newspapers became their own kind of gatekeepers about what the public should or shouldn't be exposed to. Editors and producers from the New York media industry, for instance, would meet over lunch and discuss what the scoop of the day was: Which gossip was worthy of coverage? Which politician should be platformed? What big story should be reported on? This overlapping coverage between papers and news organizations was so strong that it actually created the illusion of a universal common narrative. It was the de facto network behind the network news.

This wasn't necessarily a codified conspiracy to dictate one shared truth to the masses but instead was an emergent trend toward covering the thing that was new, novel, and easy. It made economic sense to cover your competitor's stories because that was a way to ensure that what people were talking about was the same, "best" news of the day—a guaranteed audience.

There were good things and bad things about this trend in curation and fact selection. The good thing was that there was some strong margin of consensus across the major newspapers and news networks. The consumption of news was largely the same across the population, which encouraged a sense of camaraderie and sharing in the collective narrative of the moment. People could talk about the same headlines around water coolers or the dinner table. This kind of shared consensus seemed to resemble a facsimile of objective reality, since everyone was discussing the same things. It informed a certain type of patriotism, a connection that came to define the previous century of American life.

But having vast centralization of news in the hands of few meant that the popular editorial narrative became the narrative of the majority and those with power. It was censorship by omission of minority issues and alternative perspectives.

This control of the narrative was cause for criticism, most pointedly brought by linguist Noam Chomsky, who wrote of the media's system of engineered consent. Chomsky believed that the news operated with what he called *filters*, and that "the elite domination of the media and marginalization of dissidents that results from the operation of these filters occurs so naturally that media news people, frequently operating with complete integrity and goodwill, are able to convince themselves that they choose and interpret the news 'objectively' and on the basis of professional news values." In his opinion, minority voices were unjustly left out of the conversation and debate.[12]

Journalistic Failure #4: Publicity Pandering

Throughout World War II, unchecked propaganda from all sides in the news reached a fever pitch, with every belligerent participating in a massive fight for public opinion. By the end of the war it was clear that information warfare was a powerful weapon—it could raise armies, incite violent mobs, and destabilize whole nations. After the war, a new civilian profession emerged and quickly swept across the world: public relations.

In the decades after, the PR industry became massively successful, and it infiltrated much of our shared media. This was, for the first TV news, a blatant concession to the will of powerful corporations. In the early days, this concession was extreme. Some of the earliest news programs on television were literally named after major advertisers, such as the *Camel News Caravan*, and they consisted of an anchor reading the day's headlines.[13] This was an era in which the fledgling medium of television news was magical to an audience. Those who'd never been exposed to this kind of show were inclined to take the information being presented at face value, and news broadcasts were bookended by what today we would consider ridiculous advertising propaganda.

News anchors would, for example, switch from soberly reporting the news to immediately explaining the benefits of smoking cigarettes. This muddying of advertising and journalism was an early failure of the television news industry.

US consumers of that era didn't know the difference in authority, and many simply trusted what was told to them on TV. These shows became extremely profitable for advertisers. This blatant pandering eventually gave way to skepticism from news consumers everywhere.

This pandering still happens, though it tends to be much less pronounced. As any PR agent will tell you, a number of the stories you see in major outlets were placed there, not by dint of their inherent newsworthiness, but because they pitched the story to a journalist who found it to be interesting. For the journalist, the story landed in their lap—making less work for them to do in finding a scoop. There are institutional guidelines in place to keep this from becoming too pronounced, and these stories are still corroborated with the same exacting standards found in the rest of the institution. But pandering happens, and it does still influence some of the news we see.

Journalistic Failure #5: Following Authority

In the wake of 9/11 and the horrific attacks on American soil, there was tremendous anger, sadness, and anguish in the press. After entering Afghanistan, the United States turned its attention toward Iraq. The French philosopher Paul Virilio spoke about the broad and building media support for the invasion of Iraq. He criticized the collective emotional fervor that nearly every major media institution embraced in the lead-up to public mobilization against Saddam Hussein: "The media, of the catastrophe and cataclysms that crippled the world with grief is now so vast . . . [it] paves the way for intolerance swiftly followed by revenge."[14]

Virilio was speaking to the emotional tenor of the media at that moment, suggesting that there was a broad panic about terrorism muddying our ability to see things clearly. According to him, the American media, seeking vengeance for 9/11, was misdirecting its fury.

It turned out he was right. The faulty intelligence that prompted the US invasion of Iraq ended up being one of the most catastrophic military mistakes in modern history. The fact that the mainstream American media maintained limited skepticism of the endeavor in the lead-up to war (and even at times supported it), was indicative of a major failure. I call this the *Failure of Following Authority*.

The fact that straight journalistic coverage both implicitly and explicitly supported the invasion of Iraq is a real cause for concern. As one prominent New York intellectual put it to me, "In regards to the invasion of Iraq, the *New York Times* is perhaps the most prominent purveyor of misinformation."[15] What he meant was that the *Times* (and most major American press at the time) wasn't willing to ask the hard questions of the government that might have deterred the botched, problematic invasion on the false pretenses of seeking out weapons of mass destruction.

This illustrates a critical problem with corroboration: *accepting what authorities say as fact*. As a journalist, it's much easier to take the government's official line as straight news. It reduces the liability for the reporter and the newspaper. Sometimes corroboration is difficult or impossible, as the *Times*, for example, had limited visibility into Iraq's weapons systems, or access to the same intelligence that the Bush administration supposedly had.

But there is a difference between journalistic institutions completely bowing to the pressure of the government and them facing the limits of their reporting capacity. It is another thing to equate that with sheer bullshit.

The lack of corroboration and the failure of the mainstream media to challenge the official narrative in the lead-up to war was disastrous for Iraq, Syria, and, to a far lesser degree, America itself in the decades after the invasion.

This is a fundamental failure mode for journalists: relying on authorities. In many scenarios, it's not easy to confirm or deny reports. Instead, journalists simply relay the official line of the government. This provides protection for the journalist and the paper (in case they're wrong), and is also simply easier than digging deeper.

The Ways Journalism Works

Journalism is not just a stream of failures. It actually works well because the institution is set up to check itself. As individuals, we are very bad at questioning our intuitions about what is true—especially when it comes to emotionally arresting information that we *want* to be true. As we explored in part I, we all have biases, and we must rely on others to check these biases for us to improve our sensemaking capacity. This, fortunately, is what journalism was built to do when there is a thriving ecosystem of news producers following a well-established path.

Journalistic Safeguard #1: The Correction Cycle

Let's return to the case of the Iraq invasion. The US government's efforts to build a case for war were substantial, and there wasn't much good enough information available to corroborate Colin Powell's claims that Saddam Hussein had weapons of mass destruction. Skeptical individual reporters and opinion writers didn't have enough evidence to counter the official government reports.

However, before we declare journalism morally bankrupt in response to this catastrophic failure of intelligence gathering, we know because it eventually checked itself.

We know many newspapers got it wrong only *because* of the reporting done by other journalists further down the line. Journalists examined the facts and couldn't find reason to believe that the invasion was justified. WMD were never found, and we know they weren't in Iraq because journalists spent a decade looking for them. Investigating the allegations of the US government, reporters scoured the country attempting to corroborate the claims and eventually found them to be false. News reporters kept doing their job, updating the public record and capturing attention by stoking outrage over the government's initial misinformation and mistruths.

The system worked to correct this mistake—far too slowly—but it eventually worked.

This failure came at a huge cost to the integrity of the US government. The false pretenses of the war cost $1.9 trillion, many thousands of dead civilians and military personnel, and more than nine million refugees.[16]

In a world *without* these news organizations corroborating, an alternate scenario might just as easily play out, one in which weapons were "found" by the invading United States forces (a fact still believed without evidence by many partisans). Without journalistic verification and the institutions behind it, this papering over of the blunder would have been very possible. In much the same way, authoritarian governments like North Korea declare the disastrous Korean War a victory, and Putin systematically casts doubt upon any press that disagrees with his version of events in Ukraine.

Even though we may be thoroughly disappointed with how long the process took, the system did work. This *correction cycle* is how long it takes for revisions to the accepted narrative to be published.

Journalism is "the first draft of history," as the adage goes, but that draft must be corrected over time. This means that most of the time, the first attempt at making sense of the event is going to be at least partially wrong. Just how wrong depends on many factors, including the veracity of eyewitnesses, governmental intervention, conflicting perspectives about an event, or confusion in a chaotic situation. But corrections eventually do come.

Journalistic Safeguard #2: The Swiss Cheese Defense

In an ideal world, correction happens much faster. While single papers may get a story wrong, the chance of an egregious falsehood passing through multiple papers lowers every time it's published. A single paper may report inaccurately about an event that occurred at a protest. But another organization running through the same filter and the same sourcing process may actually catch the falsehood through its own verification process.

This is what is known as a *Swiss cheese defense*. No individual news organization is 100 percent accurate, but the chance of survival of a lie transiting between two, three, or four different news organizations drops dramatically.

As the journalist Jonathan Rauch states in his 2021 book, *The Constitution of Knowledge*, with traditional journalism working correctly, a hypothetical falsehood might have a 50 percent chance of making it to print. But that same falsehood would have only a 25 percent chance of making it through a second journalistic filter of another paper, and a 12 percent chance of making it through a third paper. When more news organizations report upon the event through their own process of verification, the likelihood that a falsehood will spread to the masses decreases enormously.

This is important to recognize because within the network of news organizations reporting on topics and events, there is what Rauch calls "positive epistemic valence"—in other words, a fundamental trend toward truth. News organizations compete to scoop one another on facts, and these mutual verification processes help organizations see one another's blind spots. News organizations keep each other honest, as the loss of face in getting something wrong can cause enormous damage both to readership and to professional reputations.[17]

As a mechanism for basic sensemaking in our communities, journalism works. But it's not free. As audience attention floods to social media, so goes the capital that once powered newspapers. This shift is most palpable in the newsrooms of local newspapers, where journalistic operations are quickly disappearing.

The Downward Spiral of Local News

When my father started working at the local newspaper in early 2012, he was already an experienced technology journalist who had spent his career running a trade magazine covering tech behemoths like IBM and Microsoft.

When he joined the staff of the local paper, there were seven people on staff, which served about seven thousand people in the county where I grew up. That was half of the staff that had been there in the previous decade. When he left in 2015, there were two, he and one editor, running the entire newspaper.

In the early 2010s, many of the disruptions and layoffs that had come to define local journalism were already in full swing. A sister paper, the *Napa Valley Register*, had seen similar dramatic declines, losing more than half of its staff, dropping from a head count of thirty to fifteen. Together, they served a population of close to one hundred thousand people with their local news.[18]

Our local paper was an anchor for the community. Op-eds ran and kept a lively debate going about local school board elections. Mayoral races were actively discussed. It was widely read. I remember clearly when my sister was featured in a front-page story for building a computer in the early '90s (when such things were rare), and became a bit of a local celebrity for several weeks.

This collapse was a multiyear process. My father's paper, the *St. Helena Star*, went through this change in a five-year span as it shifted from an active newspaper reporting on local affairs to a local content outfit with much more limited scope.

In that period when my father worked there, he watched the newspaper cease some of its fundamental functions. In order to maximize profit and minimize costs, they stopped copyediting and fact-checking. By the end of his tenure, he and one other editor were literally responsible for all the news written in the paper.

This was not for want of important news. During his tenure, a large earthquake shook the Napa Valley, the largest quake to hit the Bay Area in twenty-five years. It injured hundreds, killed one person, and caused more than $350 million in property damage. Just three years later, enormous fires swept through the region, killing forty-three people and causing well over $1 billion in damage.

Events like this and the reporting that comes with them would historically have been a boon for news publishers, providing a critical service to a community trying to make sense, rebuild, and heal. The quality of my father's paper was still high, and he won an award for his reporting during this time. Another regional publication, the *Press Democrat*, won a Pulitzer for its coverage of the fires. By critical standards, it was still high-caliber reporting.

But still the cuts persisted. The business model was collapsing under the pressures for profit, along with a new, strange informational environment the newspaper was operating in. The newspaper was a business, an organism trying to survive, and if the environment for news changed and the business didn't work, it didn't really matter how good the content was.

When the fires swept through town, people didn't wait for the local papers. They went to Facebook and Twitter for up-to-date news and happenings about the disaster, relying upon citizen reporting and hearsay. Much of this information was inaccurate, but it was much faster.

Because of the internet, the incentives of the paper had changed.

Why Local News Is Dying

This breakdown in local journalism wasn't unique to the small town in which I grew up. This collapse of local news is happening all over the world as small publishers struggle to make ends meet.

You've probably heard the simplified response that "it was the internet!"

that caused the decline of local papers. The truth is more nuanced. As the media sociologist Jeremy Littau has said, the so-called Golden Age of newspapers had long ended by the 1990s.[19]

Throughout most of the twentieth century, newspapers large and small were money-making machines. They'd built reputations for strong factual reporting, and because of that they built monopolies on attention.

Throughout the '70s and '80s, subscriptions had been falling, but not enough to create substantial problems. News organizations through most of the twentieth century had become fully consolidated into chains. In the '50s and '60s, they'd become part of conglomerates, publicly traded with profit margins often reaching between 30 and 40 percent—enormous gains for any business. Investors came to expect these margins, and this profit made them handsome targets for acquisitions. Chains like Gannett, McClatchy, Knight Ridder, and Lee Enterprises (the company that purchased my father's paper) aggressively bought local papers during this period. They took on substantial debt to do so, because when the papers were extremely profitable, this kind of debt obligation was sound business logic.[20]

When the internet came along, these overleveraged assets sitting on a hundred-plus-year-old business model were in for an unpleasant wake-up.

Suddenly, these newspapers didn't have a monopoly on attention and local advertisers. First Craigslist arrived, detonating the classifieds business—which slashed as much as 40 percent of the revenue of many local papers. When Google and then Facebook came along, the vast majority of the remaining advertising spend began to jump ship as well.

The hefty debt that media conglomerates had taken on to justify their ownership in all these local papers came due. The results were deep cuts, massive layoffs, and the gutting of one of the most important mechanisms for local democracy: the regional paper. An overleveraged business model was what broke the backbone of local news.[21]

Social media hasn't yet become a viable alternative for local news. Nothing yet really has. Social media doesn't bear full responsibility for killing it—people simply went elsewhere with their attention. But its absence is still felt in our communities. Remembering what local news used to do, for our neighborhoods and our sense of place, can point us toward new possibilities. Hopefully, that memory can inform what takes its place.

IN SUM

We began this chapter learning about what journalism actually does, following the story of a young Joseph Pulitzer as he found his way into the news business, writing down his outrage about being ripped off by a swindler. By publishing it, he provided an essential service to his community in the process.

We then blurred our eyes to see that the shape of a newspaper, in its basic form, is not dissimilar to that of a social media company, only differing in size and serving instructions. Recognizing this, we can see that a handoff between journalism and social media companies should have happened, but didn't. We can begin to see how the failures of social media very much resemble problems with journalism itself.

Journalists provide distinct services to the general public that answer three primary questions:

- Corroboration—Did it really happen?
- Curation—What is important?
- Critical Opinion—Why does it matter?

Journalism is not foolproof and regularly falls prey to inherent failures that stem from its underlying incentives: *selective facts, seeking conflict where there is none, groupthink, publicity pandering,* and *following authority,* among others.

A core reason journalism works is because, despite how it presents itself, *its own authority is not infallible.* Sitting among a community of news providers, news organizations compete to scoop one another on one another's failures. This gives us helpful epistemic defenses like *the correction cycle* and the *Swiss cheese defense.*

As the number of journalistic institutions dwindles, the power of these defenses weakens, much like an immune system slowly degrades. Unfortunately, social media's rise, coupled with overleveraged debt from large news conglomerates, has eroded the institution of journalism, particularly impacting local newspapers.

As institutions and journalism fray, we lose important collective tools for making sense of the biggest challenges in our communities. In the coming chapter, we'll see what happens when we face a global pandemic armed only with our own intuitions.

Chapter 23

Trust and Truth

I have a foreboding of an America in my children's or grandchildren's time—when the United States is a service and information economy; when nearly all the key manufacturing industries have slipped away to other countries when awesome technological powers are in the hands of a very few, and no one representing the public interest can even grasp the issues; when the people have lost the ability to set their own agendas or knowledgeably question those in authority when, clutching our crystals and nervously consulting our horoscopes, our critical faculties in decline, unable to distinguish between what feels good and what's true, we slide, almost without noticing, back into superstition and darkness.

—*Carl Sagan,* The Demon Haunted World, *1995*

A Misinformation Microcosm

Truth has never been easy. Our shared truths have emerged from a long and painful struggle to find a common narrative of what constitutes "reality." And nowhere have the difficulties of our collective sensemaking been more on display in modern times than in the attempts to make sense of facts during the COVID pandemic.

In March 2020, for a brief moment, a window of consensus opened along with candid reflection on the impending threat of COVID. No battle lines had been drawn. No punditry had yet been spun. It was an odd time in which everyone across the political spectrum seemed to agree in good faith that there was a terrifying problem that needed to be solved. COVID hadn't yet been fed into the outrage machine. The pandemic hadn't yet become a moral weapon.

Of course, this didn't last long. When the machinery began to spin up, it did so quickly. First around the severity of the problem, then around the origin of the virus, then around testing, before long around masks, and finally around a persistent and frightening set of conspiracies about vaccines. We all had a firsthand view into how the confusion blossomed. In the days immediately after the beginning of COVID-19, a WhatsApp group shared among my extended community in New York was on fire. Most people in my community were just waking up to the reality that COVID was going to dramatically change the way they were living their lives.

The impending calamity was clear to me only because I know a handful of public health professionals, and follow a number of academics and researchers who study risk analysis. I knew COVID was likely going to be a fundamentally different crisis. By dint of my work and my knowledge diet, I was just a few weeks ahead of the curve. I found myself stocking up on pasta and masks nearly a month before the rush and encouraged my friends to do the same.

From my core community in New York, a pandemic-specific Whats-App splinter group quickly formed with several dozen extended friends, all concerned about COVID-19. Initially, everyone that joined was an admin of the group. That made roughly thirty of us with equal privileges to moderate. In two days, in the wake of the sudden collective panic around COVID, the group ballooned in size as friends and friends of friends joined, eventually hovering just around the maximum number of WhatsApp members—256 people.

As quickly as it grew, the quality of the knowledge shared in the group steadily began to decline. The tenor of the feed went from one in which people asked thoughtful questions to one in which people were hyperventilating in text form, spewing conspiracies and rapid-fire questions. No one was in control.

Speculation began quickly sweeping across this group. Screen captures showing message exchanges with friends who "knew someone" at City Hall who had inside information about imminent draconian government measures. Rumors of an imminent citywide lockdown of highways exiting the city, with a full militarized quarantine with police in the streets began spreading first. Then texts came about a run on the banks, and an impending shutdown of the subways, trains, and bridges.

Before this point, I believed that most of my friends and acquaintances were similar to me in how they filtered factual knowledge. I assumed that there was a formula they were using to determine what was true, a similar algorithm for reality.

I was very wrong. The group was spiraling into a paranoiac web of conspiracy. The sheer quantity of unverifiable information was overwhelming.

Attempting to parse the chaos, fighting to keep things under control, I began asking for the sources of each claim—and debunking rumors as well as I could. One morning in mid-March, I woke up, and all admin privileges had been removed by the group creator, except mine. I'd been elevated to the de facto lead admin of this group of hundreds of friends and strangers, all of whom were desperate for good information about the rapidly unfolding pandemic. I had been put in charge of keeping the garbage out.

This move to put me in charge wasn't democratic. The group's creator was a friend who had seen how much time I was spending trying to keep things accurate. He sensed that my proxy network somehow was working ever so slightly better than most in actually monitoring what was real. People suddenly looked to me to make determinations about accurate knowledge as the pandemic detonated in New York City. And as nice as it felt to be seen as an authority, it was a horrible job to have.

A Matter of Citations

As the primary admin and moderator, I was suddenly forced to build and implement some functional guidelines for content, ones that would bring this large community into line with the best facts. Fortunately, in my research for this book, I was buried deep in the history of truth finding and epistemology—the study of knowledge itself. I began looking to basic verification systems from the past.

Long ago, before Gutenberg invented the printing press, collective human knowledge was a mess of misinformation. Books, the only semipermanent repository of human knowledge, had to be painstakingly copied by a scribe or slowly printed by a hand-powered ink block. Scholars spent huge amounts of time, from months to years, trying to duplicate a single book. And since books were rare, the ability to compare multiple texts was even rarer. Because of this, corroborating the observations other scholars made about the natural world was slow and tedious. Scholarly books might have alchemical conjecture mixed in with real scientific observations. A magical incantation might sit next to an accurate astronomical measurement. For those trying to determine what was true, there were few ways to corroborate and compare observations. Much like a modern WhatsApp thread, you couldn't discern what was fact from mere speculation.

The printing press changed this by increasing the sheer quantity of texts available. Having multiple copies of a book allowed for more books to physically be in the same place at the same time. Observations and conjectures could now be compared across texts. Misinformation and mistruths, interspersed with real observations, could be corroborated or debunked for the first time. The act of citation itself became possible.

This fresh accretion of shared knowledge allowed for something like a new, permanent record of observation. A network of available citations of other works began to emerge, slowly building the foundations of shared empirical observation of the world.

Citation, the simple process of referencing an idea in another published work, is still one of the most powerful tools for gathering and objective truth we have. It's a process adopted by librarians, scientists, and journalists after centuries of combating rumors, hearsay, and viral mistruths. It's also something that only select corners of the internet have embedded in their design.

In the midst of the explosion of chaotic information, I instituted a strict rule of basic citation in our WhatsApp group; participants had to reference a *primary source* (the original publisher of the information). Only links with proper citations from known and established news sources or academic articles could be used. If there was speculation made, it needed to be backed up with evidence: a hard link to a study, a news article, or the original publication. No memes. No gossip or speculation. These were unpopular decisions that weren't made democratically, and policing them was no fun. But they worked. For a moment, the standard of forced citation calmed the turbulent ocean of conspiracies and rumors.

Conspirituality

Several months into the pandemic, when the web video *Plandemic* found its way into the thread discussion, it caused a major fissure. The slickly produced documentary, purporting to show a web of conspiracies behind COVID, was full of verifiably false statements, and it was quickly called out. A number of the second- and third-degree connections who had joined the group pushed back against the decision to discredit the film.

As a result of this and other decisions to curtail misinformation, a whole new WhatsApp group splintered off from the primary thread. The splinter group was called, "It's OK to ask questions," and it proceeded to become a clearinghouse for viral theories, unsourced ideas, and misinformation wrapped in a spiritual package. I thought most of these friends shared my protocols for sensemaking. It seemed that they should have known better.

But something else was going on. Tracking threads on social media elsewhere, I could see that this particular group wasn't alone. Public Facebook groups I was a part of were shuddering with similar convulsions of confusion. With a sinking feeling in my gut, I realized that this division and subdivision of communities into fiefdoms of knowledge, both good and bad, wasn't happening only in my personal network. This same explosion of misinformation was happening around the world on social media simultaneously—literally billions of times over as subgroups splintered into subgroups of like-minded individuals.

It's important to note that in many of these groups there were no engagement algorithms at play in this propagation of misinformation. WhatsApp had no ranking algorithm at all. It was just humans sharing what they thought was important. They were using a different kind of algorithm: a community algorithm. These were actual filter bubbles emerging in real time. And they were very bad at figuring out what was real. There was an epistemological train wreck coming, and I could see straight down the tracks.

Bad Proxies

We trust our communities with our lives. When a majority of people in our community believe something, we feel compelled to believe it, too. This explains why and how so many people refused vaccines even when their friends were dying from COVID. We tend to believe what our communities believe, even when it threatens our lives. As the psychologist Brooke Harrington has suggested, "Social death is worse than actual death."[1] Research by NYU psychologist Jay Van Bavel has shown that these belonging goals often trump accuracy goals when we're trying to make sense of the world. This is the reason truth often takes second place to the approval of our in-group. For our Paleolithic ancestors, belonging to a group of others was far more beneficial to our survival than determining accurate truths. This is how we're wired: Tribe first, truth second.[2]

We all use proxies for knowledge in our lives. We don't expect our neighbor to be able to drill a cavity out; we expect a dentist to have that knowledge, so we pay for their privileged know-how. We don't expect to be authorities on legal affairs when it comes to incorporation or a business dispute; we go to a lawyer.

But when it comes to opinions about broader facts, we have no such presumption of expertise. People who sound smart that happen to look and sound like us are often considered authorities enough.

These people are not always good proxies for capital-*T* Truth. They are, perhaps, good references for *opinions*, but they're often not complete authorities on the topic they have an opinion about.

I call these people *trust proxies*. They act as sort of epistemic anchors in our communities and online, providing us with what we presume to be privileged information. They often exist in our immediate network as people we know as being smart, well-read, or ideologically credible sources. They can also be people we follow closely online. I was acting as one of these proxies for my community.

Sensemaking as a Group

Our wiring for belonging and in-group cohesion is some of our deepest, most powerful programming. Yet because it's so deep, it's also fairly invisible, usually operating unconsciously on a daily basis despite being intricately integrated into our social worlds. We're constantly looking for belonging and approval.

This often manifests in political partisanship and a disposition to be skeptical of any disconfirming belief that goes against the grain of our party.

Viewed charitably, partisanship can be seen as a set of shortcuts that enable us to make decisions about difficult issues. To our ancestors, this was the value of our tribe. We often used the wisdom of our immediate community to make judgments about the best course of action. Our tribe provided us with moral frameworks that made it easy to answer difficult questions. They made it easy to act when faced with complexity.

It can be exhausting to live within complexity. Having to deal with the intricacies of life's tough choices is time-consuming and fundamentally difficult. We are, as researchers Susan Fiske and Shelley Taylor say, "cognitive misers." It's much easier to listen to the truths of our in-group than to expend tremendous energy on figuring the answers out for ourselves.[3]

But while we might be more accurate about many topics with the help of our tribes than we would as individuals, we're certainly not as accurate as we might be if we source the best information possible from the wider world. When these tribal narratives take too large a role, they reduce our ability to see objective reality. It takes work and energy to stay in the middle ground around issues that are complex. That complexity is lost when we only listen to the narrative of our in-group.

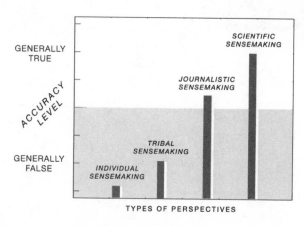

Degrees of accuracy in types of sensemaking.

Tribal biases are often better than individual abilities to determine what is accurate, but are far from the most effective models for determining truth. Journalism and scientific sensemaking fare better.

Even Our Trust in Science Is a Social Trust

Upon reflection of my time trying to parse truth from fiction during the COVID pandemic, I challenged myself to be honest about the reasons why I trusted the vaccines while many people I knew were afraid of them. The reality was that my immediate proximity to public health experts in my personal life gave me a sense of familiarity to the issues of COVID. I could call to mind experts I personally knew when addressing the problem.

That familiarity was everything. I'd like to think that it was some special powers of deduction that I was born with, but really it's just the fact that I was situated within a network of privileged information.

If I was someone living within a community that was many degrees removed from public health or governments or institutions in medicine, I would approach the vaccine and its advocates with extreme and healthy skepticism. It's only because I know these people that I feel comfortable trusting them and what they stand for. This social trust is the proxy my mind was using to decide the vaccines, and the efforts to contain the virus during the pandemic, were "safe."

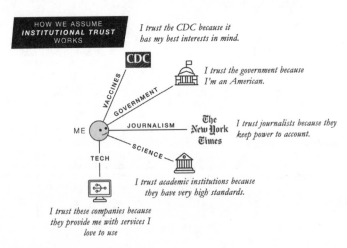

Why we think we trust institutions.

Most vaccine skeptics exist within social circles that have other vaccine skeptics. They live within clusters of normative vaccine apprehension, a type of networked skepticism. Simplified rejoinders like "trust the science" might as well be "trust the strangers" for those who live within these communities. My proximity to scientists is what established that trust.

Institutions Are Foreign Things to Our Tribal Brains

Name any institution. They *probably* provide some very important value to society. But it's unlikely you can verify that personally. Does the FDA really test vaccines for maximum safety? Does the CDC really independently verify that masks are the best way to stop a viral pandemic? For that matter, does the IRS really send all of your tax money to the federal government? Does your vote really get counted by the State Board of Elections? Unless you yourself have worked inside these institutions, you don't know for sure. Instead, you must model if you *think* they are trustworthy. You use others to determine that trust: a trust proxy.

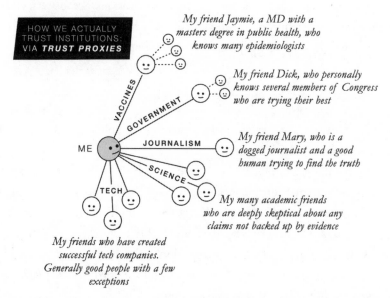

We use trust proxies to validate institutions that we have no direct access to. On social media, these can also be influencers we trust and follow, but do not know personally.

Institutions are strange, unwieldy things to our paleolithic brains. These brains want us to seek privileged information about what is actually happening inside these giant impersonal entities. We don't want to trust them, but we're forced to. We're forced to because of laws and rules that were put in place before we were born that say we need to do what they say.

But more importantly, we were forced to because we need them for society to operate. These institutions are enormous machines that allow us to get things done as a collective. The modern world society is fully dependent upon them, and placing our trust in them is a fundamental part of the broader contract we have with society. The police, the fire department, the library, the school that taught you to read. Each of these require our collective trust to run smoothly. When trust in them begins to falter, rightly or wrongly, these institutions lose power.

In this moment of profound media skepticism, it's extremely unlikely people will listen to an institutional mandate above the recommendations and norms of their community, especially if the government isn't in the hands of their political party.

Most elites feel comfortable with these institutions because they know people who work within or around them. They have an easier time trusting them because they are actually adjacent to their operation.

Institutional Aliens

The writer Scott Alexander unpacks a healthy metaphor for understanding this phenomenon, which I'll paraphrase:

Imagine if a group of aliens landed on Earth and told us that humanity has been infected by a terrible, deadly parasite. They say they didn't bring the parasite but say that only *they* can diagnose it using their advanced technology. The parasite is otherwise indistinguishable from a terrible flu, except that it will kill millions of humans if we don't act now. If the aliens are given enough time, they can develop a special new injection which, they say, might just save your life. We just need to trust them.

You don't know any of these aliens personally. They seem like they might be credible. But they're also literally aliens. They clearly have advanced technology. But are you going to trust them with the well-being of your life and your family's health? You'd be right to be skeptical.

Alexander uses this metaphor to describe what it must feel like living within a community that has no contact with the professional class of public health officials, doctors, and epidemiologists. They might as well be extraterrestrials offering an alien cure to a previously unknown disease.[4]

But it's important to remember, we *used to trust institutions.* How did social media turn us into aliens?

The Erosion of Common Truth

If COVID-19 did anything well, it was to expose the imperfect tools we normally use to figure out what is true. If the pandemic was a stress test for our epistemic abilities, it was one that we failed miserably.

In the earliest days of the pandemic, an official media consensus emerged, suggesting first that the virus originated in a wet market, in Wuhan, China, a place where wild animal meat (so-called bushmeat) is served for public consumption. Such markets have a history of being transmission points for zoonotic viruses—viruses found in bats, monkeys, and marsupials that are not known in human populations. If food preparation is done improperly (for example, a cut on a hand), these viruses can jump to humans. And since our immune systems have no knowledge of these viruses, they can rapidly explode across human populations and wreak havoc. A few noteworthy examples of this are HIV and Ebola, two of the most devastating modern pandemics, both of which very likely originated from bushmeat.

But for COVID, we had very little information. The wet-market narrative was supported by an official report from the WHO that was cited in most major news outlets. This origin story became front-page news and editorials. CNN, the *New York Times*, Fox News, and most other media reported it.[5] In the earliest days of the pandemic, almost no one had enough evidence to make a conclusion as to where the virus came from. But it was a crisis and people needed an explanation.

There were, of course, counternarratives to be found. Twitter and Facebook users, as well as many experts, offered some reasonable doubt as to the mainstream accepted story. They noted that the outbreak occurred in the same city as the Wuhan Institute of Virology, where coronaviruses are extensively studied under the highest security, biosafety level 4.

As the pandemic progressed, the possible story began to fork into many many different branches. Depending on your political persuasion and information sources, you were likely to believe a different version of the virus's origin. After a year, the Wuhan lab-leak theory suddenly became more than just a rumor; it turned out to be an officially plausible origin of COVID.[6] But what made that different from the other theories in circulation? Depending on which community you were a part of, you may have heard other pernicious narratives: that COVID had been planned as a way to shut down the global economy and allow China to ascend. COVID was a conspiracy to take down the Trump presidency. Or it was an overblown hoax. Or it was a bioengineered weapon. The difference between a wild conspiracy and a credible theory was entirely dependent upon who you followed.

Throughout COVID we felt this narrative whiplash many times. In the early days, we were told not to wear masks because medical workers needed them. This was the best available authoritative information—and also one that ended up being a false choice—one that might have actually caused the virus to spread more quickly. When the CDC backtracked, saying masks were indeed helpful, many people were left scrambling to make their own. It's possible that this *was* the best information of the moment and *did* save the lives of medical workers. But what was left was a public with a fundamental loss of faith in the system.[7]

In a crisis, we feel overwhelmed by the sheer number of competing narratives we hear and see, and we're not sure what to trust. There are so many hypotheses assaulting us on a regular basis that it's hard not to call out the false ones aggressively. We feel a strong desire to reduce the number of speculative theories flying around. We want to reduce the chaos.

In the absence of a good truth-finding system, we default to our political biases. For most of us, this begins with shutting down the narratives of our enemies, by making fun of them or castigating them. We acquire a type of skepticism born from frustration and exhaustion, making us overly suspicious of the other side's narrative, even if it's right. When critical narratives are changed, we feel justified in our anger. We direct it at those responsible for the broken narrative: the very journalists, scientists, politicians, pundits, and policymakers responsible for getting us facts. They are rightly critiqued for being incorrect, even if they are doing their best.

But this is a very hard problem. As we've become aware of the vast market-place of different perspectives in reporting, truth itself has become harder to find. The internet has birthed a universe of reporting that upon its facade resembles legitimate news. Among amateur pundits, YouTube personalities, and various social media stars, very few use the standards of journalism to report. These influencers have begun to amass enormous audiences that rival traditional media organizations. The follower count of many of these politically oriented stars on social media dwarfs those of many Pulitzer Prize–winning papers. Much of it is opinion masquerading as news, or, worse, false or deceptive reporting meant to burnish reputations and peddle bogus supplements. Not all of it is bad, of course, and a huge portion of these people are doing their very best to figure out what is real.

But I want to caution us as we begin to unmoor ourselves from the land of objective assumptions of journalism: It is a dark and problematic place, one that you can probably feel seeping into any conversations you're currently having about politics, statistics, or objective facts.

What's left is a myriad set of moral instincts that pull us toward the stories that "feel" right to us. These are the stories that don't significantly challenge our worldview, but instead play to our personal fears or biases, rather than collective systems of verification. As objective truth becomes more difficult to see, we gravitate toward the things that just feel right to us. Our superstitions become tangible. The conspiracies against our group become plausible.

There is a dangerous form of relativism here in which all assumptions are carried forward as equally valid. All claims are considered equally true. When all knowledge is flattened into unverifiable statements, we lose more than academic boorishness and the prognostications of gray TV anchors. We actually lose a semblance of shared reality.

This diversity of opinion and narrative has steadily contributed to a new, dangerous era of confusion. It is the beginning of an unintentional slide into an informational emergency, and with it a full-fledged crisis of trust.

At the end of the day, we don't have the time to do the work journalists do. We can't spend our days talking to experts or verifying sources. We don't have a better system yet for parsing what is true. Even as these old institutions contort and crumble under the strain of the internet, they still do a job no one else has yet replaced. Journalistic principles are still one of our best instruments for truth-finding, even if they are imperfect. We haven't yet figured out what should take their place.

Let's return to the giant epistemic funnel we spoke of before. This funnel was built with journalistic principles embedded in its design. A lot of information goes in (mostly garbage), and a tiny amount of knowledge comes out the other side (verified news).

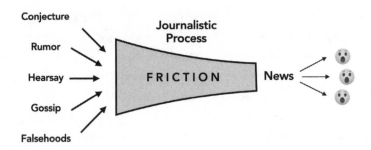

Since this process used to be something we relied upon journalists to manage in society, we now need to do it on our own. Instead of the massive, society-wide epistemic funnel, we were left with our own personal, tiny funnels of sensemaking. And right out of the box, these don't work very well.

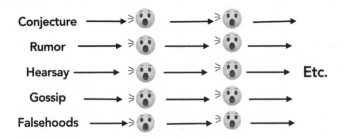

As this chaos increases, we go with our instincts. Our gut wins out. And when everyone is following their gut, we all lose.

IN SUM

We began this chapter exploring one of the hardest and strangest stress-tests of our sensemaking system: the COVID pandemic. During the outbreak,

I found myself in a difficult role as the primary moderator of an extended WhatsApp community of more than two hundred friends and strangers who were desperate for good information. This community came to resemble a microcosm of our breakdown of shared reality online. A few key features of this group challenge some common assumptions about social media:

- *It had no engagement algorithm,* but misinformation and conspiracies still flourished.
- This misinformation was curtailed only after *undemocratically implementing rules for moderation.* These were fairly simple rules for citation, based on referencing a primary source from a reputable publisher, while avoiding speculation and memes. These rules turned the chat into a valuable resource throughout the lockdowns.
- *Enforcing moderation policies made people mad enough to leave.* Gently but firmly maintaining the rules fixed much of the misinformation, but also caused at least one subgroup to splinter off into their own community where they could share conspiratorial questions on their own. This type of fragmentation into like-minded groups shows how so-called echo chambers form, through persistent self-selection.

In this chapter, we learned what happens when we lose institutional trust: We default to smaller ideological enclaves where our intuitions are not challenged. Inside these enclaves, we defer to *Trust Proxies*—individuals with perceived expertise on a topic area. These individuals become stand-ins for institutions, which gives them great power. Many online influencers now operate as these institutional stand-ins, amassing audiences that rival traditional news networks, often without real expertise.

If social media has inadvertently undermined institutional trust by collapsing shared knowledge, what has it done to our shared narratives? Next, we'll explore how one of our most cherished principles is surprisingly vulnerable to destroying itself when it meets virality: freedom of speech.

Chapter 24

Freedom of Speech vs. Defense of Truth

In the wake of Black Lives Matter protests in 2014, political commentators Jon Stewart and Bill O'Reilly convened for a debate on *The Daily Show*. The protests in Ferguson, Missouri, partly spurred by social media, had garnered enormous mainstream attention, and the issues of race and policing were heavy in the air. The topic of the conversation was white privilege, and, unsurprisingly, they did not agree.

Watching the clip now, far removed from that moment, I was struck by something peculiar about it: The two men were actually speaking in code. Through their exchange, Jon Stewart carefully, step by step, attempts to break down O'Reilly's resistance to the idea that White people are born with privilege in the United States, while O'Reilly pushes back with gusto. During the debate, Stewart pressed O'Reilly: "I want you to admit that there is such a thing as white privilege...I just want you to say that 'I'm terribly, terribly wrong on this.'" O'Reilly fired back, suggesting if there's White privilege then there must also be Asian privilege, referencing statistics that show Asians make more money than Whites.

A fiery exchange ensues, culminating in O'Reilly raising his voice, shouting, "America is now a place where if you work hard, get educated and are an honest person, you can succeed!"[1] You can almost hear the tense attachment to a narrative in O'Reilly's voice. His narrative was clear: that privilege is less important than opportunity in determining success. He was defending an important ideal that anyone can make it in America. Stewart, his voice also rising in response, replied: "If you live in a neighborhood where poverty is endemic, it's harder to work hard! It's harder to get an education." By the end of the twelve-minute debate, O'Reilly surprisingly concedes, agreeing with Stewart that white privilege is a factor. As a viewer of this exchange, I was left with a strange aftertaste, feeling like two grand narratives were fighting it out in real-time. A stage with two conflicting ideals, both seeking validation from the audience.

For cable, the debate was shockingly intelligent, and I recommend watching it. There was no ill will between them, and though they were both trying to score points from the audience, they were focusing tightly on the issue at hand—trying to make sense of a charged national issue. The men were not debating facts. Instead, they were debating two important but conflicting national narratives, and doing so in a surprisingly persuasive way.

Social media has done something very strange to our common narratives. It's pushed historically marginalized issues deep into mainstream discourse. Many, many minority issues are now openly discussed. And for better or worse, these narratives we are exposed to—as in the case of this debate between privilege and opportunity—can be highly contradictory.

The answers here aren't easy.

We feel tremendous resistance when our narratives are contested. It feels as if the fabric of our society is being put on trial. And in a way, it is. Nations consist of a shared tapestry of narratives and beliefs that allow for us to maintain a common identity. We might find ourselves defensive, angry, and disgusted when these core beliefs are challenged. Every country needs some baseline of shared narratives in order to function. Some of them must overlap with factual reality, but not nearly as often as we think. The belief, for instance, that "the US Constitution was made to adapt and work for everyone" is a shared belief that was core to the creation of the United States, but is not a requirement for every citizen to hold.

Not all of our shared narratives are correct. In fact, many of them are probably half-truths.

For example, you've probably heard the story that George Washington had wooden teeth. Suffering from terrible dental hygiene, he actually had a number of sets of teeth throughout his life. Some were a combination of ivory teeth and animal teeth, as well as several pairs that included the teeth of the enslaved. How he got these teeth isn't totally clear in the historical evidence, but archival records indicate that he paid someone for "Negro teeth." In that era, many people sold human teeth to dentists, who at the time peddled whole sets of real teeth as lead-molded dentures.[2]

This is, of course, a problematic and unpleasant image for contemporary audiences—one of the founding fathers walked around with stolen human teeth in his mouth. And this isn't the least of his behavior that can be viewed with scorn through a contemporary lens. He acknowledged rigidly enforcing strict punishments upon the people he enslaved. He brutally dealt with those who tried to escape to freedom.

But Washington was also, by most public accounts, a man of his word. By the standards of the time, he was willing to sacrifice much for the cause of the emerging American nation—putting his life on the line to try to build a resilient democratic republic. It's not difficult to identify the hypocrisy of enslaving humans while fighting to free a country from tyranny.

Which of these facts should we choose to focus on when we retell one of our country's founding stories?

If you already hold some skeptical views of Washington, more contemporary portraits of other nuanced and flawed leaders are easily found. Martin Luther King had numerous extramarital affairs that have become highlighted in recent decades. Mahatma Mohandas Gandhi espoused explicitly racist ideas as a young man in South Africa, saying that Blacks "are troublesome, very dirty and live like animals."[3]

These three powerful men had serious flaws, but many don't prefer to remember them this way. Our impulse is to explain away their transgressions where we can, or carefully balance them against the norms of the day because the ideals they represent feel valuable.

Social media is particularly good at revealing new narratives from previously hidden corners of our world. And as we've seen repeatedly throughout the book, the most controversial narratives tend to spread the farthest. When we revisit the past with so much new viral information, it's hard to keep our most precious narratives intact. Unexpectedly, this poses a real problem for one of our most important cultural values: freedom of speech.

Minority Opinions

At what point might dissatisfaction among a minority prompt a change in the majority's opinion?

Minority opinions always lose in a democracy. This is enshrined in the rules of the game: people vote, the majority wins. When I say "minority," I don't mean it in the way it's often used colloquially to describe historically underrepresented or marginalized groups. I mean it in the mathematical sense: a smaller number or part of the whole. Most struggles in a democracy come down to this primary question: When should the majority adjust their behavior to satisfy a minority?

Democratic systems require balancing the needs of the many against the needs of the few. Democracies cannot automatically concede to the needs of minority groups, just in the same way that any group of people would find themselves paralyzed and uncoordinated if every individual has different goals.

But minorities do matter, insofar as everyone will be a part of some minority at some point in time. Our beliefs are not always uniform, and at one time or another we need to figure out how to square them with the beliefs of others.

Minority voices don't matter as much as the majority's. It's possible that reading this statement has touched a nerve or captured a thread of modern politics running through your head. Discussing minority rights and opinions has become politically charged—a partisan object. But before you conclude I'm slipping into a political frame you might disagree with, let's back up and explore a real-life sampling of some contemporary minority opinions.

One: A group of people who, after getting vaccinated, have severe, life-threatening complications. They are among a minority who have real concerns based on their personal experiences. They hold legitimate anecdotal evidence in hand showing that vaccines are unsafe. They believe that no one should be vaccinated, and that vaccines are fundamentally dangerous to the world, and that corporate profit is driving their adoption. They are a small but vocal minority.

Two: A minority of people with a particular self-identified gender believe that all people should address them with unique identifiers. This group believes they should be addressed by their own unique pronouns—"zie," "zim," "zir," "zirs," and "zirself," and would like these pronouns to be legally recognized on their government documents. Zie are a small but vocal minority.

Three: A US-based cultural minority believe they have the right to keep women from holding any position of professional work outside the home, and that their children do not need any kind of education beyond their religious text. They are a small but vocal minority.

Four: A group believes the 9/11 terrorist attacks and the deaths of soldiers were God's punishment for gay and lesbian acceptance in society. They believe that the country should make homosexuality illegal. They are highly litigious and regularly pursue lawsuits against state and federal governments. This results in them winning large cash settlements defending their rights to openly picket the funerals of dead soldiers. They are a small but vocal minority.

Five: A group of neo-Nazis believe that the Holocaust was a propaganda campaign, an event that didn't really happen. They are a small but vocal minority.

Six: A group of people who have personally lost family members to fatalities caused by police officers believes that all law enforcement organizations should be completely disbanded as a solution to curb violence and discrimination. They are a small but vocal minority.

I don't share these examples to imply that they are at all similar or that they represent similarly sized groups of people. Instead, this sampling of minority opinions serves to illustrate a point: It's very unlikely that you believe that 100 percent of these groups' beliefs should realistically be entertained, and that they all should be the subject of the collective attention of your country. (If you do, then you're also in a small minority.)

The purpose of explaining these individual anecdotes is to show how quickly the adage of "respect all minorities" breaks down. If we assume a philosophy of universal inclusiveness of minority opinions, this is where we hit a brick wall.

Minority opinions are not inherently fringe. All minorities have legally enshrined rights and voices that matter. Statistical minorities, unfortunately, must share space with some of these more extreme minority groups. I want to illustrate *how* a system of collective opinion needs to balance the needs of the few with the needs of the many.

To do so, we need to break out of the current partisan political frame. The point instead should be to find methods of providing dignified and unilaterally balanced ways of including minorities' opinions to add constructively to the conversation.

Ideological extremes are rarely helpful starting points in determining how we should operate in society. Yet this is exactly where we've ended up in recent years as we've shattered the common window of discourse. Fringe opinions have become so visible that we have found ourselves debating at the edges, where almost no one is.

Confused as to what is acceptable, we're no longer speaking about practical issues; instead we find ourselves at the margins of ideological debate, unsure of how to make any real progress.

If it's not possible for us to include every voice in the conversation, how do we decide which voices to include? We need a more comprehensive language for understanding this explosion of vocal minority identities and opinions.

On Tolerance of Intolerance

Following the Charlottesville protests of 2017, when a group of neo-Nazis rallied around the removal of a statue of Stonewall Jackson, a neo-Nazi sympathizer drove his vehicle into a crowd, resulting in dozens of injuries and the death of one protestor.

The scenes were horrifying, and social media erupted. Twitter and Facebook both exploded in a cacophony of triggers and moral absolutes. A flurry of tweets expressing sympathy with the victims coupled with statements of condemnation cascaded across the internet.

One particular tweet caught my eye.

Tweet:
Your right to a belief stops where it denigrates any human. #Fucknazis

On its surface, this felt emotionally right. But rereading it, I realized it was a logical trap. It presented a philosophical dilemma: Nazis, though reprehensible, nevertheless still fit the definition of physical humans. So this Twitter user, by denigrating a human Nazi, lost their self-proclaimed right to hold this belief. The tweet ate itself; it was a contradiction.

Why does this matter? This seems rather silly to focus on one person's emotional tweet made in the heat of a moment of national reckoning on white supremacy.

But if we're building a system of rules that society is governing within, you need to account for such legal and logical fallacies; otherwise the rules just don't work. It sparked a question for me that I could not dismiss from my mind: How tolerant should we be of intolerance? When does intolerance break the system? When can we deplatform others for holding a toxic belief? When, for example, is it OK to punch a Nazi?

This is the crux of the issue facing us as we decide whom to platform, whom to promote, and whom to censor. It immediately illustrated for me just how serious a predicament we are in when it comes to freedom of speech. The First Amendment of the US Constitution's Bill of Rights was supposedly written to account for this. Free speech, by default, includes many, many weird and problematic minority opinions. And the way we protect it should be clear in the document. The First Amendment states:

> Congress shall make no law respecting an establishment of religion, or prohibiting the free exercise thereof; or abridging the freedom of speech, or of the press; or the right of the people peaceably to assemble, and to petition the Government for a redress of grievances.

The amendment's purpose in its intent was to keep *governments* from infringing upon the rights of citizens. After the amendments adoption, a number of thinkers tried to build a broader philosophical case for its value to society. Most notably, John Sutart Mill, one of the fathers of modern liberal thought.[4]

The code had a purpose in its intent. But it shows the inherent conflict in a large multifaceted society with many opinions. What exactly should be tolerated? If an ideology is intolerant of others, when should you be so outraged as to shut them down, punish the holders of these opinions, and/or censor them?

In the eyes of the First Amendment, minority opinions cannot be seen as uniformly toxic or righteous. The result of this, in principle, was to provide equal footing to bad ideas and good ideas so they could compete, allowing for the maximum possibility of the best ideas to win out.

But it's hard to feel like some of these crazy ideas are valuable in any way, particularly when they seem to threaten many of the people and concepts we care about. Let's pause for a moment and think about the difficulties of being a member of a tiny fractional minority. It takes a significant measure of empathy to understand how you would feel if put in a situation like this, of holding a deeply unpopular minority opinion. At one point the effort to abolish slavery was an unpopular minority opinion in large portions of the United States. The complete censorship of these views could have extended the inhumane practice even longer than it lasted. Obviously, comparing a Nazi's opinion to an abolitionist's opinion isn't the same thing to anyone standing near the mainstream, but in the eyes of the law, both are protected speech.

The reason is that *even bad ideas have value*. Here Mill's argument goes like this: When you're trying to organize a lot of humans, censoring individuals—beyond just being painful for the individual—robs *society* writ large of the opportunity to correct the idea. There is wisdom in criticism and correction. An open debate and repudiation of a Nazi sympathizer benefits society more than a quietly held bigoted belief.

Mill stated in *On Liberty*, "If all mankind minus one, were of one opinion, and only one person were of the contrary opinion, mankind would be no more justified in silencing that one person, than he, if he had the power, would be justified in silencing mankind."[5]

Mill believed that criticism advances truth. Through conflicting opinions, we end up with better understanding, better tolerance, and better ideas.

He who knows only his own side of the case knows little of that. His reasons may be good, and no one may have been able to refute them.

But if he is equally unable to refute the reasons on the opposite side, if he does not so much as know what they are, he has no ground for preferring either opinion.[6]

So in Mill's view, toxic ideas *must* be allowed in order to allow for them to be refuted. If we stop people from having the opportunity to learn through criticism exactly why an idea is wrong, then we rob people of education itself.

So, if liberal tolerance is a fundamental value that helps us move forward as a society, when is too much tolerance a bad thing? When are we supposed to stop tolerating the things that are offensive, or intolerant of others? How much of this toxicity must we accept, and amplify into our common discourse with social media? When does being over-tolerant break the system?

The Paradox of Tolerance

The philosopher Karl Popper referred to this dilemma as the Paradox of Tolerance. The paradox goes as follows: "Unlimited tolerance must lead to the disappearance of tolerance. If we extend unlimited tolerance even to those who are intolerant, if we are not prepared to defend a tolerant society against the onslaught of the intolerant, then the tolerant will be destroyed, and tolerance with them."[7]

The key here is how we define tolerance and harm. And how we define that needs to be based on a common lexicon of what it means to be *harmed*.

It is not as easy as it sounds to agree, as the definition of harm is a slippery one. To use the previous example, the South in the lead-up to the Civil War used the "way of life" argument to defend slavery. Abolishing slavery would "harm" their way of life, even as it literally forced millions of other humans to live in slavery.

The more contemporary philosopher John Rawls, in an attempt to define this more clearly, wrote: "Society should be structured so that the greatest possible amount of liberty is given to its members, limited only by the notion that the liberty of any one member shall not infringe upon that of any other member."[8]

Is this belief, if realized, a threat to the liberty of any other member of society? Here too, it gets slippery, as the collectively adjusted moral frame we use to assess the "other members" of society is often varied, arbitrary, and deeply inconsistent. We rely upon authorities, who hypothetically share our values, to determine where the line of harm is and who constitutes a real member of society.

As mentioned, enslaved people and women historically were not considered real members of society. Dehumanization, or subhumanization, is a terribly easy way of reducing people below a level of rights. A clever lawyer of the era might argue they weren't harmed by not having rights because they weren't really "legitimate" citizens. This is, of course, by our standards moral nonsense, yet it still stood up for years in courts and public perception.

Another frame can be drawn from an old saying in law that reads something like this: "Your liberty to swing your fist ends just where my nose begins." The exact source of the quote is unclear, but it is said to have originated with Oliver Wendell Holmes, John Stuart Mill, or maybe even Abraham Lincoln. According to the saying, your right to exercise your liberties ends when those liberties endanger the life and safety of another individual.

But there are two sides to this: the side of the person swinging the fist, and the side of the person who perceives the harm. Clearly there are bad things we can do to people that don't just include punching them. Change the parameters of the *perception* of harm, and suddenly this becomes painfully murky. As we have watched in recent years, perception has been fundamentally skewed by social media, leaving an entire generation unsure of what harmful speech really is. Speech is not a fist, but it can feel like one when amplified online.

This is Not a New Issue

In 1807, President-Elect Thomas Jefferson, in his inaugural address, famously said the following in reference to people whose speech might undermine the nation: "Let them stand undisturbed as monuments of the safety with which error of opinion may be tolerated, where reason is left free to combat it."[9]

Before he became president, he was a free-speech maximalist. He literally had a preference for newspapers over the government. He felt that people could largely self-organize around problems using the news as a guide and manage their own society without state authority. Newspapers and the Constitution were all that was needed for people to get together and solve the problems of the day. Report a problem, solve it using the deliberative body of government. That was his ideal.

But once he was president, this ideal didn't last. He ended up doing an about-face on his original sentiment later in his presidency when he instructed the state attorney generals in New England to prosecute the editors of various newspapers for saying things he disagreed with. When his philosophy about free speech met the realities of governing, his philosophy buckled.[10]

The struggle over the issues of speech, moderation, and censorship have been at the center of American politics from the very beginning.

We Can't Expect Censorship to Be on Our Side All the Time

One of the core reasons free speech is so important, is that the act of censorship itself is not on the side of any particular ideology.

Politicians and judges have throughout history interpreted the First Amendment differently. Many have read it in moral and emotional ways, and made calls that today we would call excessive and unjust.

For example, in 1835 the Postmaster General effectively censored anti-slavery materials sent out by the Anti-Slavery Society of New York by stopping the flow of letters and refusing to intervene when pro-slavery protestors seized and destroyed the mail.[11]

Today, we struggle with these same issues. Republicans express tremendous anger at the perceived censorship of conservative voices on social media platforms, and worry about the liberal values of employees influencing major policy decisions. This anger comes at the same moment that conservatives are passing laws to ban certain books in libraries that disagree with their moral worldview. This happens just as liberals are looking to increase the prevalence of hate-speech restrictions across the board, in the midst of a rapidly expanding definition of what hate speech actually is.[12]

Every era of the past has been wrong about some facet of the truth. This era is no different. Truth is found when ideas are allowed to collide, to be debated, to be proven out, or to fail on their own. By following these two simple rules, we can get better at finding the truth from generation to generation:

1. Speak in good faith, and let others do the same.
2. Do not suppress others' speech, even if you disagree with it.

These are hard rules to follow as individuals and in society, but we should do our best to adhere to them whenever we can.

These guidelines are morally ambiguous by design. The rules should not inherently bias us toward one particular partisan set of ideas. Criticism pushes truth forward. It makes us smarter. As Jonathan Rauch says, "We can much more easily see the faults in others' arguments and ideas, and we kill our hypotheses instead of each other."[13]

The Real World

Let's keep these principles in mind and apply them to the real world we live in today.

No one's speech should be censored by the government. The First Amendment protects these rights. The rules for platforms, however, will continue to be tricky and contentious. The default, of course, should always be less censorship whenever possible. Instead we should address the issue of *amplification*. Someone, somewhere, will always make the determination as to what information can and cannot be spread far and wide. For over two centuries, these people were media gatekeepers. Today they are largely content moderation algorithms. As we've learned, these algorithms, in conjunction with media incentives, orient our attention toward extreme content. In practice, this dynamic amplifies more radical ideas and fringe minority opinions.

This inherently puts good-faith, constructive debate on unequal footing with extreme perspectives. Modern social media actually amplifies bad-faith interpretations of most issues. It pays people to find the worst versions of ideas, giving prestige and reach to those who are best at being outraged.

Criticism on social media is not meant to advance understanding but to score points. Grandstanding has come to dominate most debates, and truth is often outcompeted by falsehood.

Freedom of speech is not the same thing as amplification. As my colleague Renée DiResta put it, paraphrasing Aza Raskin: "Freedom of speech isn't the same thing as freedom of reach."

This recognition is helpful when we consider the emerging threats of technologies that allow for the issues of content moderation will become even more prevelant. Computational propaganda, coupled with chatbots that are indistinguishable from humans and wide-open viral amplification, will make all of these problems worse.

IN SUM

We began this chapter with a debate between two pundits, Jon Stewart and Bill O'Reilly. While they acted as though they were debating facts, they were actually debating two broader structural narratives: the American dream of self-empowerment, and the enduring disenfranchisement of Black Americans. This debate was a valuable exchange between two minds presenting their best arguments, and helped viewers come to a better understanding of the issues.

Social media has made debates like this far less common. It has increased our exposure to new and contentious minority opinions that challenge shared narratives, without giving us the ability to debate them in good faith.

Minority opinions are valuable when they can be effectively debated, and they are protected in the Constitution through the First Amendment. But not all minority opinions should be platformed. While the First Amendment protects minority opinions from *government* censorship, it does not account for virality, or the reach of those ideas. The right to amplify or demote content still lies with the owners of platforms themselves. Currently, contentious and extreme opinions tend to receive the most engagement online, outcompeting thoughtful commentary and reasonable discourse. While it feels new, this is actually a tricky problem that has been at the center of American politics since the founding of the republic.

If this sounds depressing, it need not be. I am optimistic that this is a solvable problem. In the following chapters, we'll begin exploring some principles of good design that might help improve good-faith dialog and discourse in our online spaces. (A list of these broader design principles can be found at outragemachine.org.) We can keep the core idea of free speech intact while building better incentives for constructive debate, prioritizing and amplifying the conversations and criticism that advance our understanding of the world.

As we enter into part V, we'll explore how these issues scale upward into a much broader set of problems for democracies writ large. From this higher vantage point, we'll start to see some solutions emerge.

We'll begin with a parable about a tiny island in the middle of the ocean, and an unexpected visitor who changed it forever.

PART V

Rewiring the Machine

Chapter 25

The Parable of the Island

Not long ago there was a small island nation in the middle of the ocean. It was inhabited by five farmers, who each owned a roughly equal part of the land. The island was many, many miles across—big enough that its few inhabitants didn't ever see one another unless they made plans far in advance.

This island had elections. Every few years, the five farmers got together and formed a small council and voted upon who would be the executive of the island and represent it to the rest of the world. This job was important because it set up trade deals with the mainland and managed the long road that went all around the island. A simple majority agreed upon the executive. If no one won a majority, the election happened again every month until they came to a winner.

But the island was large enough that it was almost impossible to tell if an executive was doing a good job. So they hired a young man to do reports for the island. Every week, this Reporting Man went all over the island and wrote up a paper detailing how things were going and sent it to each of the farmers. The Reporting Man had a hard job, but it was valuable because each farmer was busy with their own land. They paid him a small sum for this service.

The current executive of the island was a very capable woman who did her job well. But she and the Reporting Man didn't always have a smooth relationship. The Reporting Man was always on her case, telling everyone else what was wrong on the island, which often made more work for the executive, who was already very busy. But the executive needed the Reporting Man's support to win the election, so she talked with him often, explaining what she was doing for the island. Everyone was generally fine with how the arrangement worked.

One year, one of the farmers, a rather noisy farmer they called the Loud Man, was out in his field digging a very deep well, when a puddle of oil began bubbling out of the ground. Watching the black liquid rise from the mud, he realized his fortunes had just changed—he could make a lot of money from this oil. But the oil was almost certainly not just on his land—it was probably all over the island. He knew he'd have to share it with the other farmers if they found out.

So he kept the oil a secret and hatched a plan. He would try to become the executive at the next election so he could sell the oil rights and make himself rich. He began wooing the Reporting Man, treating him very nicely, while secretly plotting against the executive.

Before the next election, he began telling the Reporting Man how bad the current executive was. But he had a problem: The executive wasn't so bad, and there wasn't that much to complain about. So he began making things up. "I heard the executive is padding her pockets," he said to the other farmers at their annual meeting. "I think she's corrupt."

The Reporting Man began to investigate. But he couldn't find any evidence of what Loud Man was saying, so he put that in his report. "Loud Man's accusations are not correct," he wrote. The Loud Man, realizing he was exposed, demurred. "Well, I still think something's fishy here!" he said.

At the next election, Loud Man pounded his fist on the table and made a big stink out of things, wanting to be the executive. But the other farmers were unconvinced. According to the Reporting Man, the executive was still doing a good job. So no one really believed Loud Man's bluster, and the election came and went without an upset. The executive continued her hard work, and things went along as they always had on the island for a few years.

One day, a salesperson arrived on the island in a boat. On the boat he had a big antenna, which he offered to place in the middle of the territory, giving everyone the ability to send messages. The antenna was linked up to a little transmitter that could be used anywhere to take pictures and share words with everyone else on the island. This sounded like a good idea, so the farmers pooled their money and bought the antenna.

The new antenna was great. Each farmer could report on things they saw that were wrong with the island and share it with each other and the executive. The executive handled these problems, and the island got a little better, bit by bit. Everyone used the antenna, and the Reporting Man loved it because he could track everyone's problems and use it to write up faster reports. It made his job a little easier.

But no one loved the antenna more than the Loud Man. He would use the antenna to share his opinions about *everything*, not least of which his opinion on how he could do a better job than the executive. The other farmers mostly still ignored him, and the Reporting Man corrected the Loud Man wherever he could.

One day later that year, an enormous storm hit the island. There was severe damage to the road connecting the farmers, and the Reporting Man suddenly had his hands full trying to keep track of all the things that were being done to fix it.

During this hectic week, the Reporting Man botched one of his reports. He wrote that a broken part of the road had been fixed, but it hadn't been fixed yet. He wasn't perfect, and had gotten the repair schedule mixed up. It was a big island, and he couldn't be everywhere all at once. *No big deal*, he thought. *I'll just put it in my next report with a correction. That's what I always do.*

But the Loud Man caught wind of this, and saw his opening. He used the antenna to post pictures of the broken road. "Look! The Reporting Man lied! There is something unusual going on here!" he said. "I think the Reporting Man is in cahoots with the executive! You cannot trust them."

The Reporting Man was embarrassed. He had never had his job challenged like this before. He sent out an apology, and said he would do better. This kind of mix-up happened sometimes. But it was never a problem before.

But the Loud Man didn't hold back. "You can't trust that Reporting Man. He has a hidden agenda. I think he's telling lies." The Reporting Man was shocked and confused. *I'm doing my best. Why am I the bad guy now?* he thought.

The other farmers were all now confused, too. The Loud Man was right. The Reporting Man had made a mistake. "Where there's smoke, there's fire," one farmer said as he stopped paying for the Reporting Man.

The Loud Man kept up his attacks and peeled support away from the Reporting Man and the executive bit by bit.

When the next election came around, there was a lot of confusion. Nobody knew whom to trust. Two farmers, throwing their hands up, decided not to vote—they just didn't know what to think with all these accusations flying around. The Loud Man, full of bluster, proclaimed, "I can do better. These guys are liars. Vote for me."

With the two abstentions, Loud Man won the election with a slim majority: one vote for the incumbent, two abstentions, and two votes for him.

Loud Man had prevailed. As the executive, he could now sell the oil rights. Knowing that the Reporting Man was the only thing now between him and success, he focused his scorn. "It's time to clean this place up! The Reporting Man is our enemy," he said. "And we really don't need him if we have this antenna." Soon the Reporting Man was out of a job.

The Reporting Man was left with no way to pay for his services and left the island to find work elsewhere. Without him, the rest of the farmers didn't really know what was happening and were forced to trust whatever the Loud Man said on the antenna.

Within a year, Loud Man had sold the oil rights in secret. Now very wealthy, he used the money to hire his own Reporting Man, one who did whatever he said, along with some muscle to make sure the other farmers voted for him every election.

The island was different. It had a new stable government, and it was unlikely to change for many years to come.

IN SUM

In this chapter we introduced a parable about a tiny stable democracy. Democracies are coordination games that depend on good information to operate. The media provides reliable information, acting like a referee that keeps ambitious parties honest. When the media is undermined by new distribution technology, the public will suddenly struggle to make educated decisions about whom to support and whom to trust. This is known as a *coordination trap*.

The enemy of a democracy is not necessarily an ambitious loud man. The enemy is actually confusion. Confusion itself is what keeps the participants from knowing who the demagogue really is. As Steve Bannon said, referring to his political strategy, "The real opposition is the media. And the way to deal with them is to flood the zone with shit."[1]

We can see this same dynamic playing out around the world in places where established media is competing with social media. Demagogues are able to stoke and sustain confusion. More confusion means fewer people can coordinate. More accusations flying around means fewer people have any idea who is actually telling the truth. Next, we'll learn about what this strategy means for liberal democracies around the world.

Chapter 26

What's at Stake

Today, liberal democracy is in danger.

Within our bubbles of contempt, we're barely willing to face the future together. Small issues are looked at as mortal threats, existential risks are seen as unimportant. Collectively, we are unable to agree upon which is which. This is a direct result of the way social networks are designed today. They are fundamentally changing the way we see truth, the way we measure our friends, and who we see as enemies.

When we project into the future, this does not resolve well. We experience real outrage on a near-daily basis, and we feel real fear about those with opposite political agendas. Among those highly engaged in politics, 70 percent of Democrats and 62 percent of Republicans say they are afraid of the other party. This is priming us to ignore the other side: the portion of Americans who express consistently conservative or consistently liberal opinions has doubled over the past two decades. The ideological overlap between the two parties has diminished dramatically: Today, 92 percent of Republicans are to the right of the median Democrat, and 94 percent of Democrats are to the left of the median Republican.[1]

As social media forces us further into our corners of anxiety and outrage, we see our worst political and social fears realized. We feel something akin to desperation, and we become more willing to look for solutions and leaders in extreme places.

This trend is reflected in worrying polls about the acceptability of an authoritarian state. In the United States, there is a sizable and growing percentage of the population that are fine to completely end democratic governance and default to army rule, as long as it's on their side. The youngest of us, who are widely thought to be the most liberal and idealistic, are not immune to this authoritarian trend. Nearly a full quarter of Americans under twenty-four years of age have said an authoritarian from their party is preferable to democratic elections.[2]

We are losing the capacity to cooperate around large goals; we're all angry about different things. Partisan gridlock in Congress is only getting worse. This is a direct reflection of public sentiment.

Liberal democracies have historically had the best repositories of ideas, an advantage because they are drawing from a diverse and unique set of perspectives to solve problems. Social media is stripping the gears of that machinery and authoritarians around the world are taking advantage of it.

A highly polarized society cannot maintain multiple parties and viewpoints. What's more, a subset of the population is particularly prone to authoritarian triggering, in which they—by default—will support an autocrat if exposed to the right stimulus.

The Social Contract

The fall of the Berlin Wall was one of the most significant geopolitical events of my life, yet when it happened it didn't mean much to me. I remember the images on CNN of protesters sitting on top of the graffiti-covered concrete barrier. Some struck it with hammers and crowbars, and some carried pieces of it home with them. I was eight years old and had no deep understanding of what the big deal was. It looked like people were having a wild party. For millennials like myself, the collapse of the Soviet Union feels like ancient history, a hazy milestone from another era.

For my parents' generation, however, it was as if the axis of the globe had finally righted itself. It marked the end of decades of a violent battle of ideas that brought our species to the edge of existential ruin. Through it, we saw the proliferation of civilization-ending nuclear weapons, the fear of which rightly defined the previous fifty years. In the clash between Western democracy and the many different emergent versions of communism, most significantly the Soviet Union, the model of liberal democratic governance had won.

For this reason, the reunification of Berlin and the collapse of the Soviet Union was a triumphant moment for the West. It was touted as proof that liberal democracy was indeed a superior way of organizing humans. The scholar Francis Fukuyama, with his landmark book *The End of History*, argued the debate had been settled. He contended that liberal democracy has consistently proven to be a more effective system (from an ethical, a political, and an economic standpoint) than any alternative since the French Revolution. Democracies would largely be the stable, most efficient way of organizing humans going forward. The problem had been solved. Western liberal democracy was the final form of human government; the apex of the evolution of mankind's political ideology.[3]

At the time, evidence seemed to support his view. The end of the Cold War defined a new era of global cooperation. Studies show that the end of the Cold War and the subsequent rise in the number of liberal democracies were accompanied by a sharp decline in total war, interstate wars, ethnic wars, revolutions, and refugees. Real-world observations showed that interstate warfare had largely ceased to exist in South America, Southeast Asia, and Eastern Europe. After their transitions from military dictatorship to democracies, these countries simply stopped starting wars. Across the board, when nations became democratic, they no longer wanted to bear the cost of violence against their neighbors.[4]

In short order, and just in time for the '90s, the consensus of the era indeed seemed to coalesce around the idea that, at least as far as political systems were concerned, we had figured it out. Governments would inevitably trend toward economic liberalism, representative leadership, and stability. We could as a species move on to other problems.

The End of the End

There were critics, of course. They suggested his theory was largely based on conflicting definitions of war and "mature democracy." Critics also argued that it was hard to generalize about democracy because it originated only relatively recently in human history. In their view, we simply haven't clocked enough run time on the democratic experiment.

So it was that in 2016 history suddenly started back up. Something strange had happened to the way that democracies operate—an unseen force that caused things, in short order, to go haywire. Populist revolutions reverberated across the world. Demagogues emerged victorious in a number of major elections. Suddenly, liberal democracies were on a tenuous footing, just a few decades after the "triumph" over other all versions of governing.

Two democratically legitimate elections—Trump winning the presidency, and Brexit—brought this home more than any others. Something very strange had happened, but what?

Our lives are entirely built around relationships, like those with our families, employers, and communities. But one of the most subtle and critical relationships we have is with our government. It determines a lot about how we live our lives, even though we don't think about it often. As we've read, social media has dramatically contributed to that relationship changing. These tools are creating new dependencies and incentives that are challenging the historical foundations of democratic government as we know it. They are reinforcing a way of looking at the world that might be fundamentally undermining the democratic project.

One of the most useful frames for understanding this breakdown is the concept of a *social contract*. The idea was an attempt by a generation of Enlightenment-era thinkers to make a case for the authority of government and natural human freedoms.

Jean-Jacques Rousseau, the most successful thinker to outline the equation, saw it like this: We as individuals choose to give up a certain portion of our basic freedoms, and in exchange the government must ensure our rights.[5]

THE SOCIAL CONTRACT

GIVE UP FREEDOMS

Individuals   Government

GET RIGHTS

This was supported by the thinker John Locke, whose concept of the social contract was so influential to the founding fathers of the United States that it was written almost verbatim into the Declaration of Independence. It outlined how "to secure these rights, Governments are instituted among Men, deriving their just powers from the consent of the governed."[6] The idea of the social contract is a core part of liberal democracy as we know it.

But if this idea of a social contract was revolutionary at the time, it was not wholly new. The first version was advanced a century before by the philosopher Thomas Hobbes, who believed in a different kind of contract, one that did not ensure rights, but security. Without this social contract, he argued, humankind would exist in a savage and violent world. He famously called this the "State of Nature," in which life was "solitary, poor, nasty, brutish, and short."

Unlike Locke, Hobbes used this argument to make the case for what he called a *Leviathan*, or absolute sovereign ruler who could ensure our safety. The cost? Our individual freedoms. He believed, after observing the discordant violence that occurred in revolutions and wars of his time, that monarchs and other supreme leaders with absolute power were far more beneficial to humanity. Without such leaders, "the condition of man," he wrote, "is a war of everyone against everyone."[7]

Core to his argument was a story about fear—fear of inevitable violence if humans are left to themselves without a ruler. The clearer he could draw a picture of an unruled man's innate leaning toward anarchy and chaos, the stronger was his argument for an absolute monarch. The fear of that natural violence is what made his case.

Before this moment in history, kings, queens and emperors drew their power from another source: God. Working with the thinking of their day, Hobbes and Locke were trying to outline a case for government beyond this "Divine Right of Kings," the premise that all kings were ordained by God, and if you disobeyed them you would literally go to hell. From antiquity onward, this linkage with God was the primary justification for authoritarianism.

After the Protestant Reformation fragmented the church, the logic of this divine right had become less persuasive to most people. (Which God ordained a king, a Protestant one or a Catholic one?) Hobbes was doing his best to shore up a case for civil society based upon the precedent of what had worked in the past. In his opinion, an absolute ruler was that system.

Hobbes was tapping into a resonant human impulse—the desire for law and order when faced with chaos. The more aware we are of threats to our well-being, the more we gravitate toward a strong and authoritative ruler. The more insecure we are, the more we believe we require a strict and powerful head of state. Hobbes believed that this existential fear was the most important force in shaping civil society.

In modern times, Hobbes's assertion has been reflected in the research of political scientist Karen Stenner. She has shown that roughly 18–20 percent of the population are predisposed to a certain kind of threat-response that makes them immediately receptive to messages of authority. She calls these people *natural authoritarians*, humans with an automatic ingrained response to so-called normative threats. When they feel threatened, or are fearful of a certain type of cultural outsider, or sense a threat to the perceived order of things, a type of "trigger" is pulled in their psyche. Once this trigger is pulled, they will default to supporting draconian policies and authorities. Stenner's research shows that, "regrettably, nothing is more certain to increase expression of their latent predispositions than the likes of multicultural education, bilingual policies, and non-assimilation."[8]

Stenner's research suggests that Hobbes's instinct was correct: When things are scary and confusing, we default to the strong voice that speaks to our personal morals.

And as many politicians have learned since then, this power relationship still holds true today: When we feel that the government isn't doing enough to keep us safe—from outsiders, disease, crime, violence, or chaos—an authoritarian message resonates best. The more distinct our fear, the more justified the message of a strongman becomes. As we become more afraid, more polarized, more despondent about the security we hold dear, the more we yearn for a different kind of social contract—a Hobbesian one.

What Democracies Are Losing

But what might we be losing with this fear-based contract? Democracies around the world at least still *seem* to be expressing the will of the people. The 2016 elections that brought us Brexit and Trump were the result of free and fair elections. Around the world, we see similar trends: Democracies doing strange things, electing strongmen and defaulting to nationalism with the open consent of voters. The evidence is in plain view on a global scale. Strange alternate versions of democracy are beginning to emerge from Hungary, Brazil, the Philippines, and elsewhere.[9]

What we're losing is a key conceptual part of democracy that is hard to reference without first stripping it of partisanship. That concept is known as *classical liberalism*. If you ask most Americans on the street what liberalism is, most will draw from the political definition we hear all the time. Liberalism, they might say, is what liberals are—the opposite of conservatives?

But classical liberalism is different. At its nonpartisan core, it is a combination of a set of principles upon which the foundation of modern democracy is built, a philosophy that dictates the organization of government. Traditionally, classical liberalism is meant to infer "liberty" and not the politics of the left. For the sake of simplicity, it's helpful to simply think about liberalism as a combination of three fundamental ideas:

- *Pluralism.* The idea that many diverging viewpoints ultimately help us make better decisions. Those differences of opinions, in good faith, actually allow for us to address problems in unique and novel ways as a society. That out of many identities and perspectives, we become one better whole. "E Pluribus Unum"—out of many, one.

- *Free Will.* Also called agency, or the idea that all humans have inherent freedom in determining the path of our lives. That we should be allowed to make decisions about our property and our pursuits in accordance with our desires.
- *Human Rights.* The idea that humans are born with inherent rights. These rights aren't apportioned selectively by the powerful but are applied equally from birth. We are guaranteed these rights by giving up some of our freedoms to the government.

If you're reading this book, it's likely that you have come to cherish these ideas and even taken them for granted. We don't often think about the kind of historical privileges that come from living in a pluralistic and free society that supports human rights: We've been allowed to practice any religion. We've been allowed to speak our mind without fear of government reprisal. We've been allowed to choose where we live, what we buy, and who we buy it from. We've been allowed to pick our leaders. We've established laws against discrimination by race, creed, ability, sexual orientation, and color. The United States, after fits and starts, began to codify these principles into laws and norms that reflect this legacy of liberalism. Over the last century, it has set a strong—if imperfect—example as it exported them to the rest of the world.

Yet there is a consensus among those who study politics that these ideas are no longer flourishing. Many political scientists are even declaring that the Western liberal democratic world order is withering before our eyes. It's not hard to see the broken treaties, the upended relationships, and the shattered norms in our international politics. The consensus is that the global democratic project is struggling.

As we face the confusion wrought by social media, what is at stake is more than our data, our privacy, or our friendships. This is actually a fight for a way of life.

IN SUM

We began this chapter unpacking how democracy is in danger. As we're exposed to more threats, more fear, and more outrage, we've become increasingly intolerant and increasingly confused. This has undermined our shared trust. A growing number of Americans are willing to support an authoritarian government as long as it shares their politics.

We're stuck in an outrage machine designed to make us feel unsafe. When we feel our safety is threatened, an authoritarian yearning can take root.

We explored the concept of the social contract, an implicit contract we have with our government. After generations of progress under Locke's social contract based on ensuring our rights, democracies have turned a corner and begun operating under Hobbes's social contract based on securing our safety.

Free and fair elections increasingly elect strongmen and yield divisive policies. Without a shared truth and way to make sense of facts, stable democracy as we've known it for the last century isn't assured.

Up next, we'll explore how democracy itself is supposed to operate like a sort of outrage machine, and we'll uncover what happens when its system for processing outrage gets overloaded.

Chapter 27

The Machine Called Democracy

Where you see wrong or inequality or injustice, speak out, because this is your country. This is your democracy. Make it. Protect it. Pass it on.

— Thurgood Marshall

Democracy is a device that ensures we shall be governed no better than we deserve.

— George Bernard Shaw

The Code for Building a Self-Governing Republic

Looking back to the inception of the United States, the framers who were responsible for writing the US Constitution were not dissimilar to engineers trying to design the blueprints of a new machine. In a way they were programmers searching for the best operating code for society. They did enormous amounts of research on the history of failed republics, read widely, and debated fiercely on the most effective way to build a representative government that would stand for years to come. They were also keenly aware of the failures that happen when mobs of people try to make important decisions. They understood the tenets of group psychology as well as anyone in that era could. The founders were iterating upon many other failed attempts, not least of which were the English Civil Wars, which saw years of violence and revolution follow many failed attempts to balance parliamentary self-governance and the autocratic rule of monarchs.

These ambitious Enlightenment-era engineers thought they could do better. The Constitution they wrote was essentially code that was pushed into the public sphere, a set of rules a group of people adopted to effectively cohabitate together.

This code—the US Constitution—wasn't written and compiled in Objective C or Python, but in English. The "machine" that ran (and still runs) this code is the collective group of humans who inhabit a landmass in North America. In ratifying and running the code, this group of people became the United States of America.

Take a moment to zoom out and think about what humans do. As hypersocial animals, we require rules and norms to live together. The larger the groups, the more specific the rules and norms needed to cohabitate. In this way, we can think about the US Constitution as a set of instructions, or algorithms, written to help us thrive: the most basic operating code for a democratic republic.

The point of the Constitution was to help these humans do a few things: most notably, govern themselves without a king. This hadn't really been done before, and certainly not in any kind of stable or durable way.

The democracy they created is a sort of outrage machine: It was built to turn outrages into policy through a clear system of rules meant to help people fix what they see as broken in the world around them.

The expanded instructions for this code were written into a series of manuals known as the *Federalist Papers*. These eighty-five essays outline how to apply the code to society and how to interpret it. They were written in just under six months by Alexander Hamilton, James Madison, and John Jay.

They knew that what they were doing was a monumental task and unlikely to succeed. Core to this effort was understanding the previous failures of every previous democracy. In *Federalist* No. 55, they wrote:

In all very numerous assemblies, of whatever characters composed, passion never fails to wrest the scepter from reason. Had every Athenian citizen been a Socrates, every Athenian assembly would still have been a mob.[1]

Essentially, even if everyone was an ideal citizen, getting together without a good set of rules for governing, the system would devolve into disorder. They knew that democracies were prone to mob rule, and that when passion and outrages spread too quickly, governments often collapsed. This was a tricky balance, and one that would, in the decades immediately after, prove itself out during the French Revolution's many convulsions into brutal violence. Specifically, they knew that demagogues were likely to ride the passions of this mob to positions of leadership, after which they would claim power and become despots.

In *Federalist* No. 10, James Madison wrote about his fear of the power of "faction," by which he meant strong partisanship or group interest that "inflamed [men] with mutual animosity" and made them forget about the common good. He thought that the vastness of the United States might offer some protection from the ravages of factionalism, because it would be hard for anyone to spread outrage over such a large distance and capture the public's passions. Madison presumed that factious or divisive leaders "may kindle a flame within their particular States, but will be unable to spread a general conflagration through the other States."[2]

The Constitution they wrote included mechanisms to slow things down, let passions cool, and encourage reflection and deliberation. They instituted a system of checks and balances, splitting the power into three branches responsible for different things, and made it just hard enough to pass laws so that the ambitions of any individual branch could counteract those of the others.

The Moral Error in the Code

This Constitution was a flawed document, written by flawed men. It was ratified only after a large number of concessions were embedded in its design to encourage the former thirteen colonies to come together and run a new program.

Many of the compromises included entrenched rules about representation that disenfranchised huge numbers of people living there, including

Black people (written to only account for three-fifths of a person, whose votes went to their White neighbors) and women, who were completely disenfranchised from having a say in how the code was run. These concessions for slavery were terrible bugs in the code that would push the whole system to the brink seventy years later.

The worst bug in the code was written into it from the start. The ratification of the first version of the Constitution made huge concessions to a deeply racist set of institutions that sought to maintain the economic exploitation of enslaved Blacks that undergirded the entirety of the Southern economy at that time. This was a fault in the code which was contradictory both in principle and in practice. As the abolitionist Frederick Douglass stated, "Liberty and slavery—opposite as heaven and hell—are both in the Constitution." He noted that the Constitution was inherently in conflict; the document was "at war with itself."[3]

This same fundamental inner conflict was present in each of the framers who were also enslavers. Madison and Jefferson both struggled with the central paradoxes of writing a document that emphasized the inalienable rights of humans while personally excluding enslaved people from those rights and keeping them in bondage. These contradictions show the deep failure of foresight of the founders.

But we can acknowledge their failings while also acknowledging that the code they created ended up being very durable because it was deliberately built for self-correction. With the Constitution, they were, as Angela Glover Blackwell has said, "punching far above their moral weight."[4]

Their system was fortunately written with a very intentional mechanism for updates: amendments. It was designed to be improved.

Pushing Updates

The biggest advantage of the code the founders wrote was that it could be revised and reinterpreted by successive generations of humans. Politicians, legal scholars, and judges from local courts all the way up to the Supreme Court do their best to look at this original codebase and its operating manual to determine what will work for the machine today.

The system had to go through several new revisions to run smoothly. The Bill of Rights was the first update to be pushed out and adopted. It included revisions to the code to account for inconsistencies and problems with the original codebase—the first ten amendments.

This first revision was meant to account for the rules for information sharing. Recognizing that the whole machine was dependent upon each individual user having access to the best possible information to make choices about what was happening elsewhere in the system. The First Amendment established what is known as freedom of speech and freedom of the press, and was meant to act like a formula for maximizing the quantity of available inputs into the system.

The founders didn't know it at the time, but they had stumbled upon an algorithm—one that wouldn't be written out in code for another two hundred years. The algorithm is known as Branch and Bound, and it was explicitly coded for the first time in the 1960s by researchers at the London School

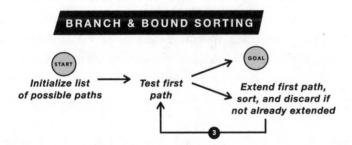

The algorithm explores branches of the pathway, which represent subsets of the possible solution set. Before identifying the solution, the branch is checked against bounds on the optimal solution, and is discarded if it cannot produce a better solution than the best one found so far by the algorithm.

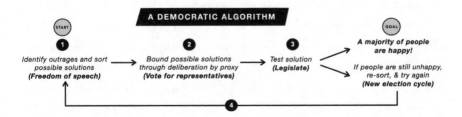

of Economics. The algorithm is a solution-finding formula, developed as a resolution to the traveling salesman problem, a class of problems that involve routing possible paths to a goal in the most efficient way possible. How do you code a computer to figure out the best path? This type of combinatorial optimization solution, as it is known, can be applied to any number of problems in mathematics and decision-making.[5]

The democratic code runs like this: If you see, hear of, or read about something that outrages you, speak to your representative. If your representative doesn't do what you want, organize and vote him out of office. If the next representative doesn't do what you want, run for office yourself. If you win, you then add your voice to the deliberative body that's responsible for writing the laws that matter to you. It is meant to help people turn their anger into debate and compromise rather than violence.

The Speed of Outrage

The original developers also knew that this machine of the United States ran on a network, the postal system. It was established during the Second Continental Congress, which appointed Benjamin Franklin as the first postmaster. This network was so critical to the operation of the machine that it was written into a clause in the Constitution when it was finally ratified in 1789. The maximum clock speed of information traveling within the system at that moment was incredibly slow, about 55 mph—the maximum speed of a horse—but averaged much slower. It was as fast as the machine could run at that moment.

In computer science, when upgrading the speed of a computer, there is a specific class of problems that emerge when machines begin clocking at a higher speed. Certain parts of the system, now running much more quickly, can overtake other systems that were running at the old speed. These are called *race conditions*, like a relay race executed out of sync, and if they change, they can cause huge problems for computing systems. As individuals we're familiar with race conditions in systems like finance: If your automatic withdrawal gets out of sync with your direct deposit, you'll get an overdraft fee.

The way that outrage transits a democratic system reveals a similar problem. Outrage has a half-life. When people get upset about a thing, their anger is often catalyzed into action. If outrages spread too quickly, or go unaddressed for too long, they might feed simmering grievances. The unique feature of the Constitution was that it allowed for outrages to be channeled into productive outcomes, petitioning members of Congress to act on behalf of these grievances, systematically passing laws that could fix what was perceived as broken. In this way, the Constitution was written to be an effective advocacy mechanism, adapting to the needs and frustrations of its population, and turning them into laws.

The key was ensuring that there was *accurate* information being shared within the system, that perceived injustices are real and not fake. Otherwise, the system overcorrects (by passing draconian or extreme laws) or undercorrects (by ignoring the simmering frustrations of the populace), both of which can cause further sparking of grievances.

There was another chunk of code, protected by the First Amendment, that became essential to helping the machine run. That code was journalism: a competitive system for verifying true and false information, supported by the public's curiosity, interest, and engagement. They used the code of the First Amendment as a foundation to build a mutually beneficial information-processing system. Modern newspapers came online at about the same moment as the United States started growing rapidly. Newspapers began to provide their service system-wide. After a few fits and starts toward the end of the nineteenth century, this new code did a decent job at filtering garbage information out of the system.

The developers were fortunate, because the code actually stayed modestly stable for the first fifty years. But when the clock speed was upgraded to the speed of the telegraph—that is, near instantaneous transmission across the continent—things began to fray. As the historian Daniel Crofts wrote "The acute phase of North-South sectionalism coincided with the arrival of the telegraph, which rocketed information ahead at speeds that seemed miraculous."[6]

Without good verification of information, and when lies and spurious claims could transit unchecked, many things started to break down. When people could spread information without corroborating it, things got out of control.

System Failure

The increased speed of information transmission through the new telegraph, coupled with an inherently problematic set of racist laws, was a recipe for a terrible outcome. Newspapers were unable to maintain their fact-checking of spurious claims, and this caused the system to fully break down in the 1860s. President Lincoln, forced to reckon with the moral inconsistency of slavery written into the original document, watched as his election brought the full and complete succession of the Southern states. Large parts of this Civil War were sparked by a failure of journalism to effectively invalidate the conspiratorial and ridiculous assertions of the Southern states. Early journalism was not up to the task, and the nation broke apart.

As Annika Neklason wrote for *The Atlantic*: "In the months leading up to the Civil War, fear festered in southern living rooms and legislative chambers. Newspapers reported that the newly elected president, Abraham Lincoln, held a 'hatred of the South and its institutions [that would] cause him to use all the power at hand to destroy our country' and that his vice president, Hannibal Hamlin, was not only sympathetic to the plight of black Americans but was himself part black."[7]

These types of commonplace stories were false narratives that profoundly influenced the political direction of the Southern states. Spurious rumors supported by false reports about Northern plots to incite revolts among the enslaved ultimately caused the machine to split in two.

The Civil War was a devastatingly violent reset of the system, one that cost hundreds of thousands of lives, killing roughly one in every fifty Americans and maiming many more.

This bloody reboot paved the way for amendments to rectify several of the most egregious original errors pertaining to civil rights and slavery. It included passage of the Thirteenth Amendment and the establishment of many new rules as to what the federal government could do and enforce, most notably that slavery was illegal and fundamentally wrong.

It's hard to imagine things working out differently, and many historians believe that civil war was an inevitable part of resetting a broken system and reinterpreting the original code of the Constitution. But it is arguable that it didn't need to get to that point. Moral reckonings were built into the system. Outrage operating at the right speed and pace might have allowed for the machine to adjust.

The code underpinning America's democracy had a goal: Help a group of diverse humans, spread over an enormous landmass, come together as one people and address their outrages in a constructive way. This is how the machine is meant to operate, and how it might need to operate into the future in order to succeed. The machine is a tool for self-determination of a people. It's meant to help reduce the suffering of the largest number of its inhabitants and help them collectively make sense of the world as best they can.

IN SUM

In this chapter we unpacked a broad metaphor for how a democracy works: operating like a machine running code written in its constitution.

Using this machine, constituents observe problems in their midst, become outraged, and through a process of deliberation and compromise, turn these sentiments into legislation directed at resolving them. We learned that the machine has conditional speeds at which it must metabolize outrage. If it responds too quickly, it might *overcorrect* for outrages by passing draconian laws that cause further problems. If it responds too slowly, it might *undercorrect*, leaving simmering grievances unresolved. We can trace the origins of major pieces of legislation to the outrages that preceded them.

New technologies that upgrade the speed of information often also come with increases in misinformation. These periods can put the whole system at risk of failure until they are rebalanced by verification and moderation processes (like journalism) that allow for the most egregious falsehoods to be kept in check.

Next, as we round the corner to the penultimate chapter of this book, we'll learn where our outrages might end up doing the most good by exploring a hidden force that shapes much of human behavior, regardless of our desires.

Chapter 28

Where Should We Place Our Outrage?

I am dragged along by a strange new force. Desire and reason are pulling in different directions. I see the right way and approve it, but follow the wrong.

—*Ovid*, Metamorphoses I

Why do people do bad things?

So far in this book we've explored the systematic origins of many of our current outrages. While this is important, the last thing I want to do is encourage a sense of moral relativism. I don't want you to quash your anger and disgust. Good and bad are not just arbitrary and fully subjective value judgements. People do bad things, and those bad things deserve punishment.

In the midst of such intense moments of cultural disgust and frustration, there is a powerful impulse to point fingers at our enemies and wish them harm. It's as if evilness has sprung forth from people's souls unheeded, and must be burned away with outrage and cancellation. If only they could be held accountable for the damage they cause the world, then we would feel a sense of justice!

But punishment alone will not solve our problems. The language of cause and effect is not present in our condemnation of others. We are missing a social conversation about the origins of this bad behavior. Whole groups of people are usually not fundamentally bad. Humans are a part of systems, and systems are sometimes designed in ways that make people do terrible things.

Nazi Germany, for example, was the result of a number of bad actors implementing a horrific system and exploiting the populace. It was not a country categorically full of depraved sociopaths. The structures and incentives built into the Nazi regime's institutions encouraged people to act in terrible ways. If we understand the factors that cause people to do awful things, we can get closer to avoiding horrific outcomes.

I don't want you to walk away from this book thinking that outrage is inherently wrong. Instead, I want you to walk away understanding that people usually do bad things because of bad incentives.

An Ancient Demon Written into Code

In 1956, the poet Allen Ginsberg wrote a poem called *Howl*. It begins with what is perhaps one of the most striking openings of any modern work of poetry.

I saw the best minds of my generation destroyed by madness, starving hysterical naked...

In it he described a faceless god named Moloch, a biblical deity of child sacrifice. Moloch is an amorphous but powerful evil force, something resembling capitalism, greed, industrialization—something—infecting humanity with its terrible designs.

Moloch whose mind is pure machinery! Moloch whose blood is running money! Moloch whose fingers are ten armies! Moloch whose breast is a cannibal dynamo! Moloch whose ear is a smoking tomb!
Moloch whose eyes are a thousand blind windows! Moloch whose skyscrapers stand in the long streets like endless Jehovahs! Moloch whose factories dream and croak in the fog! Moloch whose smoke-stacks and antennae crown the cities![1]

I remembered the poem vividly, but it wasn't until the process of writing this book that I was sent an article written by the psychiatrist Scott Alexander exploring this poem in the context of game theory. His influential 2014 piece, "Meditations on Moloch,"[2] outlines what he sees as the real "force" of this fictional deity Moloch manifesting in society. Moloch is *a system of bad incentives.*

He lays out these concepts in a series of hypothetical stories. Each one illustrates a strange force that drives human behavior. He calls these forces *multipolar traps*, or *coordination traps*.

Put simply, these traps are things that individuals and groups of people do, even when they might know it's the wrong thing to do.

Examining human behavior, you can find many examples of coordination traps—some awful, and some benign:

Imagine you're at a casual concert and everyone is sitting down with a clear view of the stage. At some point, a single person near the front stands to get a slightly better view. Those behind them must now stand in order to see. This cascades and suddenly, before you know it, everyone needs to stand in order to see the stage. Almost no one is better off, but everyone needs to do it in order to maintain their perspective. This is a coordination trap resulting in a worse outcome for everyone.

Another example: People keep their money in banks. If a single person begins a rumor that the bank is insolvent and tells other depositors that their money is unsafe, they can create a bank run—a cascade in which everyone fearing the loss of their money rushes to withdraw it at once. The bank, on financially solid footing before, suddenly loses its functional assets and becomes unable to operate. The *false idea* of insolvency resulted in real insolvency for everyone. A feedback loop based on a false premise created a coordination trap.

And a third, adapted from his article: Imagine a group of businesses, say a collection of fisheries, operating around a lake. Each individual fisherman takes their share of fish from the lake with no problems. But one day, they realize the lake is becoming polluted by all the fishing boats. As a result the boat owners get together, and every fisherman is asked to buy a filtration system and install it on their boats. The filtration costs something, a few hundred dollars per fisherman per month. The problem is solved and the lake is cleaned, but the cost of fish goes up as a result. All the fishermen of the lake continue to prosper, with slightly more expensive fish.

But one day, one fisherman decides he doesn't want to use the filter, and because of this, can undercut the price of fish being sold at the market. He can sell it for less because he doesn't pay for the filter. He begins outcompeting the other fisherman, soon becoming the most profitable fisherman on the lake. Before long, a few other fishermen see this and decide to also stop using the filter in order to compete again. This cascades downward until only a handful of people around the lake are still using filters. Eventually, the lake becomes too polluted to fish. This is another coordination trap, the result of which everyone is worse off.[3]

The important thing to understand about coordination traps is that they often operate *despite human values and agency.*

- The person at the concert sitting halfway back in the theater knows that it's dumb for everyone to stand up, but they will do it anyway because the person who stood up in front of them has blocked their view. Otherwise they'll miss the show.
- If you're a person who knows that the bank run is unfounded, but enough people are running on the bank, you will still try to withdraw your money knowing you're likely to lose all your money if you wait too long.
- If you're a fisherman trying to make ends meet, and fewer and fewer of your peers are filtering the lake, it makes sense to stop using the filter. There is a point in which you are unfairly carrying the burden of filtering the lake and putting your business at risk. It doesn't make economic sense to keep doing it. You cash in on the remaining fish while you still can, until the lake is polluted and barren of fish.

We can think of these as mechanistic incentives: things that operate without explicit human decision-making. When the incentives are misaligned, your personal preference doesn't really matter. *The system is pushing the behavior.* If you're paid enough to do something that sacrifices your personal morals, most people will reconfigure their moral compass, and often begrudgingly do the thing everyone else is doing.

It's critical to recognize that these coordination traps happen all the time and can even be found in other species. When trapping crabs, for instance, once you've captured several of them, you can place them all in a bucket. The sides of the bucket need not be particularly high and any individual crab can climb out on its own. But the crabs will never succeed in escaping the bucket. Every time a single crab tries to reach the lip of their prison, the other crabs in the bucket will pull them back down. No individual crab can escape because they cannot coordinate together.

The study of these coordination traps is the discipline of behavioral economics and game theory. In game theory, one of the most famous coordination traps is called the *prisoner's dilemma*, an imaginary scenario illustrating how two totally rational individuals might not cooperate even if it seems like it's in their best interest to do so.

Coordination traps like the prisoner's dilemma are part of a whole battery of scenarios studied in game theory. When people try to design effective systems—from board games, to traffic rules, to democracies—they need to think about incentives that cause people to act badly. The rules often dictate people's behavior. The bad behavior is set far in advance by flawed policies, poor incentives, and dysfunctional systems.

Scott Alexander, referencing the poem, called the strange and perverse desires associated with coordination traps by the name *Moloch*. I have adopted this, and whenever I see systems operating badly because of dumb incentives, I don't shake my fist at the people perpetuating the idiocy. I shake my fist first at the ancient biblical god of child sacrifice Moloch—shorthand for a poorly designed apparatus that causes us to throw our most precious objects into its maw.

Social Media's Demons

Once you're familiar with how to look for it, you can see Moloch everywhere on social media.

Social media is a coordination trap on a massive scale. Often, if you *don't* do the bad thing, you're at a disadvantage to those who do. If you don't say something divisive, you are often outcompeted in the attention marketplace by someone who does. If you don't accrue followers and attention by using moral and emotional language, you're at a disadvantage to those who do.

But these types of coordination traps go one layer deeper. Say one social media company decides *not* to algorithmically amplify divisive content. They may be outcompeted by one that does. Friendster and Myspace had no news feed. They didn't adapt quickly enough to the competitive pressures of the social media marketplace. Today they are irrelevant. Facebook, intensely watching competition from Twitter, copied the news feed and its behavioral dynamics. Today, Meta is in a battle to compete with TikTok by expanding their engagement algorithm and tweaking incentives for performance beyond friends and people you follow in order to maximize time on site.

What values is this reinforcing? What guideposts are in place to keep us from defaulting to our basest impulses? How do we avoid the coordination trap of bad incentives dictating market outcomes?

Let's go one step further. As these platforms become drivers of new cultural norms of attention, democracies begin to reflect them directly. Politicians in democracies are *required* to use social media in order to speak to their constituents. This makes them beholden to social media's game mechanics. Thus, a politician who doesn't stoke outrage will lose to the one who does.

How do we resolve these traps of bad incentives?

The answer to coordination traps is to have an authority step in. We need an entity to keep Moloch at bay, a coordination mechanism that helps all the good actors work together.

For the concertgoer, it is a person onstage asking everyone to please stay seated through the performance. For bank runs it's the government: The Banking Acts of 1933 and 1935 created the Federal Deposit Insurance Corporation (FDIC). In response to successive and catastrophic bank runs of the Great Depression. It guaranteed depositors wouldn't lose money, even if a bank becomes insolvent. It all but eliminated bank runs in the United States. For the fisheries, it's a fishing association with authority to punish those who don't comply with filtering the lake.

Defeating Moloch is possible if conditions for trust can emerge between the good actors. This is well illustrated in "The Evolution of Trust," a playable visualization of the iterated prisoner's dilemma by the designer Nicky Case. Based on the groundbreaking work of Robert Axelrod at the University of Michigan, it shows the principles in play to mathematically improve outcomes for all players, and to beat Moloch. These constitute three things which allow for trust and cooperation to "evolve."[4]

1. Repeat interactions—Trust keeps a relationship going, but you need the knowledge of possible future repeat interactions before trust can evolve.
2. Possible win-wins—You must be playing a non-zero-sum game where it's at least possible for both players to be better off.
3. Low Miscommunication—If the level of miscommunication is too high, trust breaks down. But when there's a little bit of miscommunication, it pays to be more forgiving.

You may see, immediately, how these three elements might be missing in social media. You also may begin to see how these tools could be designed better—specifically to account for these breakdowns in trust and cooperation.

IN SUM

In this chapter we explored how outrage isn't inherently wrong. When people do bad things, sometimes those things deserve punishment. But punishment alone will not solve our problems. The driving force behind bad behavior is often bad incentives.

We explored a helpful metaphorical depiction of these bad incentives: *Moloch*, a biblical god of child sacrifice.

This system of bad incentives creates coordination traps in which people behave counter to their best intentions. We can see Moloch all over social media. In a system that incentivizes bad behavior, everyone from content creators to the platforms themselves must participate in the machine or risk being outcompeted by everyone else.

How do we resolve these traps of bad incentives? There are a number of effective strategies that can help people coordinate better together and avoid mutually detrimental outcomes. Sometimes it's as simple as establishing better trust and communication. Sometimes it's an authority–a group, institution, or indvidual–that can help all parties break free of bad incentives. To keep Moloch at bay, we need an entity that can enforce a coordination mechanism to help good actors work together. Successful coordination requires trust, which we can build with repeat interactions, possible win-wins, and low miscommunication. Today's outrage machine poses a threat to the fundamental elements of trust, but comprehending these elements provides a plan of action for designing better solutions.

In our final chapter, I'll share what we can do ourselves to improve our relationship with the outrage machine. We'll close with sketches of solutions—ways that we might begin designing these tools to serve us, rather than the other way around.

Chapter 29

What You Can Do

May you know that though the storm might rage,
Not a hair of your head will be harmed.
                    —*John O'Donohue, "For Suffering"*

It has taken us the entire book to explain why we are in this difficult moment. But while we have focused on many of the macroscopic trends, the historical context, and pernicious forces influencing us, there is, fortunately, much we can do to change it. High-level solutions require a mix of high-level thinking and a very narrow understanding of the problems. We must look to points of maximum leverage—the places where we can make the most impact.

More specifically, the problems need to be looked at symptomatically, not generally. We all want silver bullets, but complex systems require very specific solutions. It's much more realistic to focus on single solutions rather than system-wide ones.

Let's return to our analogy of our collective information system operating like a human body. Just like the human body, when we have an infection, we need to take specific antibiotics in order to counteract it. We don't assume that curing our athlete's foot is going to cure our heart disease.

It's tempting to go further and further up the chain of systems-level problems and try to address the biggest ones—complexity issues sometimes require complex solutions. But this is the nature of solving difficult problems: one thing at a time.

How We Can Disagree Better, and Use Our Outrage for Good

Outrage is not inherently bad. The moral emotions we feel when we observe something is wrong in the world are critically important to a functioning society: They are the force that helps us mobilize and solve problems when things are broken. Channeled into the right vehicle, outrages can change the world for the better. Many of the most effective social movements in history—the US civil rights movement, Gandhi's campaign for an independent India, the women's suffrage movement, the gay rights movement—were catalyzed by outrage channeled into focused, systemic action.

Outrage is a problem when it manifests into toxic outrage—when it shuts down debate, the ability to discuss an issue constructively with the opposite side, and cascades into violence. When the system of discourse itself is threatened by our moral anger, that's when it needs to be addressed. That's when it becomes far more dangerous to society.

Similarly, social media is not intrinsically harmful. It has tremendous power to do good, as when it brings to light previously hidden harms and gives voice to previously powerless communities. Every new communication technology brings a range of constructive and destructive effects, and over time, ways are found to improve the balance.

As individuals, we can help. There are a handful of things we can do as humans to detoxify our relationship with online outrage, and our interactions with it online. Each of these are specific solutions that can help us reclaim portions of emotional agency that we have lost in recent years. Remember that this isn't just for us: by reducing our participation in the broken system of outrage profiteering and manipulation, we are actually helping reduce the overall levels of toxicity that exist in the world today—the stuff that our friends, family, and neighbors all feel.

And better, by selectively participating in the issues that speak to the best parts of us, we are increasing the effectiveness of our actions by staying focused on what really matters—opportunities for healthy outrage and mobilization in the future.

Reduce Your Triggers

Today there are infinite channels through which to source overwhelming outrage-inducing stimulus. Online is where we usually find ourselves emotionally triggered the most. If you find yourself emotionally triggered by what you consume online, there's a straightforward solution: Limit the amount of time you spend with it.

As you're scrolling, ask yourself these questions:

- Is what I'm looking at something I can use to actionably improve my life or the lives of the people around me?
- Is this something that affects people I know personally?
- Is this a real problem, or might it actually be disproportionately covered?
- Do I find myself regretting the time I spend using this service?

If the answer to any of these questions after regular use is a clear no, try a simple experiment: Delete the app for a week. Use a content blocker on your phone and computer to keep you from automatically returning to the website. Chances are you'll feel better, and find yourself with significantly more free time. The world will go on.

If you can't limit that specific time (because you use social media at work, for example), then make a concerted effort to reduce your exposure to the worst stimulus within these feeds. Train your algorithms by aggressively unfollowing and blocking specific accounts that share/evoke the kinds of outrageous content that regularly cause you to regret your time. Most algorithms allow for you to select "I'd like to see less of this" or "not interested" options beneath content you don't like. Your brain and body will thank you later, allowing you to focus your energy where it matters most.

Rules for News

Let's unpack several mental models that might help us to more accurately, and more sanely, consume the news. These are four ideas, developed by the late statistician Hans Rosling and adapted from his excellent book *Factfulness*, that can help us understand the world more completely when we're feeling overwhelmed by news.[1]

1. **More news does not mean more suffering.** Just because something is covered doesn't mean there's a proportional increase in its prevalence. Usually, it simply means that a certain story has found an audience— people who will watch it, click on it, or share it. We covered this machinery when we explored the power of metrics and optimization, and what happened to the internet beginning in 2009. The mere fact that we hear more about a particular tragedy doesn't necessarily imply that it's an epidemic.

2. **Good news is not news.** It's been tried before by many a journalist and TV producer. They have become tired of the negative bias that our news tends to skew toward. They have an idea: They'll cover meaningful stories about positive change, true improvements that involve a feel-good outcome! A story about a sizable philanthropic gift! A story about a quiet Samaritan making a difference in a dozen lives through regular hard work. Invariably, these stories are positively reviewed, but cannot compete with bigger headlines. They are relegated to local news, where they don't get the ratings news organizations need to maintain the attention of a larger audience. In general, you will not hear about good news because it doesn't sell.

3. **Gradual improvements are not news.** News that shows steady-state increases in some of the most important human development numbers, such as reductions in poverty, child mortality, or illiteracy, often don't get coverage. These stories are unwieldy and aren't easily packaged into news that news consumers want. This type of slow, methodical improvement is the untold story of human progress, but it is rarely, if ever, covered by journalists in a meaningful way.

4. **Our news diet matters.** Many things that we need for survival are likely to kill us in excess. Consuming too much water or food can be toxic to our bodies. Overconsumption of dramatic news is probably harmful to us in much the same way. We likely have an optimal dosage, something that producers of news would not like us to know.

Understanding that news operates in ways that sometimes makes things seem worse than they are should not make us ignore the news altogether. We still need to read the news in order to be good citizens participating in democracy. There are a few simple things you can do to make sure you're still getting your news fix, without succumbing to the more harmful parts of the outrage machine.

First, change your default to straight-news sources like Reuters and the Associated Press. Nearly every major news outlet from Fox News to the *New York Times* pays some combination of Thompson-Reuters and the AP for news. The AP is a nonprofit, which reduces incentives for emotionally repackaging content. Reuters has strict principles it follows for ensuring minimal bias. Both have extremely clear guidelines and standards for reporting factual events without significant opinion or commentary, making them safe bets for straight news about what's happening in the world.

Secondly, if you're interested in news from diverse political viewpoints, these services provide digests from multiple sides of the political spectrum, clearly calling out bias with good-faith takes: The Flip Side (TheFlipSide.io), Tangle (readtangle.com), AllSides (allsides.com), and Ground News (ground.news).

Finally, when possible, subscribe to local news. While often a little basic, supporting a local paper is a straightforward way to ensure that some reporters are paid to keep local government accountable.

Disagree Better, Not Less

Social media is, writ large, a terrible place to disagree with people. Many of our best, most empathetic and prosocial human interactions are hidden by this digital environment. We cannot see the looks on people's faces when we hurt their feelings and they cannot see ours. We are incentivized to publicly perform to our audience of followers, not for the purpose of finding truth or an amicable middle ground. Our sociometer—the visible public likes, comments, and shares overlaid directly upon our posts and responses—keeps us from healthy and effective discourse. When you're disagreeing, take the disagreement into a private, offline conversation.

As these contentious issues become more common, learning these prin-

ciples of effective disagreement can help defuse potential moral explosions as
they begin to happen:

- *Don't insult the beliefs of others.* This will actually often cause people to
  become more extreme in their views. Focus on the issues at hand.
- *Don't assume the motives of other people.* Understand the phenomena
  of *pluralistic ignorance* (when all parties assume something that is
  not true in unspoken moments). "They're not talking to me because
  they're an X…" has become more pronounced as the dominant narratives of social media entered our lives. Overcome that impulse and try
  to assume the best intentions first.
- *Get curious.* Investigate why they feel a certain way, seeking out the
  moral foundations we covered in part II. Try to identify the "elephant" in their argument, recognizing that they are likely operating
  from a different foundation. Try to figure out which one they may be
  operating from: care, fairness, loyalty, sanctity, authority, or liberty.
- *Share your values.* If you have the patience, explain *why* these things
  are important to you using their own foundations as a lens. If you are
  a progressive who cares about losing access to abortion speaking to a
  conservative, frame it as a *liberty* you are afraid of losing. If you are a
  conservative who cares about border security, you might explain it in
  terms of *fairness* to those immigrants who are already here.[2] Oftentimes just the process of investigating the values of those you're arguing with has the benefit of helping everyone move beyond polemics
  of angry mainstream narratives. You'll be much more persuasive, and
  you will feel better about the interaction afterward.

Lead with Love

In the wake of the 2016 election, Americans experienced the Great Unfriend-
ing: One in six Americans lost a friend due to opinions expressed online during that period.[3] These weren't just shallow friendships—they were largely
real human connections. COVID-19 has only made this worse.

One of the most insidious things about the way social media has infiltrated our lives is how it has made our *external* narratives about people more important than *personal* narratives. A personal narrative is based on an independent relationship you have developed over time with someone in your life. An external narrative is one that has been superimposed upon your relationship through media you both consume. While these can be difficult to thread apart, it's far from impossible. If the relationship is really important to you and still seems ideologically charged, consider taking a breath and steering conversations away from politics entirely. This can be done simply by acknowledging their concerns, and changing the subject to something you both care about.

The version of your friends you see online is not necessarily an accurate reflection of who they are. If you feel the emotional distance from them expanding, or feel yourself pulling away, try to speak with them in person, or send them a small note of kindness. Leading with this type of action can spark a positive feedback loop that can heal these divisions and help overcome the most personal casualties of the outrage machine.

Algorithms Are the Problem and the Solution

It's hard to understate the intensity of the problem we are facing. We are in the process of ceding control of much of our attention to the new masters of our digital information systems: nonhuman algorithms.

We've reached a point at which the majority of the information we consume is being touched by code before it reaches our minds. Whether it's a news article about something that has already gotten traction online, or a piece of content that has been packaged to go viral—most of the topics we've discussed are already part of this ecosystem dominated by these tools.

But what about private messaging platforms? How can algorithms actually be the problem when there are no algorithms?

Some of the strongest criticism of the "algorithms are the problem" thesis can be illustrated by the troubles associated with some of the simplest group messaging services. Messaging tools like WhatsApp, Telegram, and Signal do not even use engagement algorithms. They just consist of groups of people who connect online and serve each other information in chronological order.

They are the same rough version of ancient IRC chat rooms, or AOL instant messaging tools that have been available since the 1990s.

They also happen to be hotbeds of conspiracy theories, vitriol, and viral misinformation. But with basic chronological content, what algorithm is to blame for these inherent amplification problems? Unfortunately, the problem is us.

Remember, *You* Are Running Algorithms

Your behavior also follows rules. We as individuals also share information based on a set of internal predefined habits and norms for conducting ourselves.

We are complex, nuanced creatures who certainly don't like to be compared to machines. But our behavior, particularly how we share information, *does follow rules*. Since *algorithm* is just a really fancy word for a rules-based process, we should call these behaviors personal algorithms. Some of your strongest personal algorithms are biases—biases for and against certain politics, policies, and positions.

In this way, the members of a specific WhatsApp community are running algorithms themselves—these are human sorters and curators of memes, news, and conspiracies. If these individuals have poor epistemic filters, and develop a conspiracy theory and share it onward, they are in effect acting like a conspiracy algorithm. If this conspiracy theorist has one hundred thousand followers or sits in a WhatsApp group with hundreds of people, it doesn't matter if there is no intelligent feed-ranking algorithm. The conspiracy theorist *becomes* the feed-ranking algorithm.

As we self-select into communities that further determine our relationships with information, they are increasingly becoming our real-life news feeds.

As we've discussed, problematic information shared widely in these spaces tends to be "fast" shares. Things people click on and pass along without any friction or forethought.

In the near future, we will need to reckon with the fact that we ourselves are running imperfect algorithms adapted to an old information environment—that of centralized network news.

This is not to reduce the profound beauty of the human experience, or a reduction in our human-ness and unique power and insights of our intuitions. But the way we parse true vs. false information is its own type of sensemaking algorithm that must be updated to operate into our uncertain future.

As we ourselves are forced to carry the burden of epistemic processing, it is important to recognize that the internal algorithms we sometimes use to understand the world might be imperfect.

We're going to defer more and more of our sensemaking to these tools into the future (as journalists already do with Twitter, and then feed it to us). I believe clearly defining the interlocking parts of this relationship, and getting visibility into the algorithms themselves, will help us get closer to managing the enormous disruptions ahead. Otherwise we really are deferring the responsibility of maintaining liberal democracy to Twitter, Facebook, Tik-Tok, and a handful of other platforms.

A Better Sensemaking Algorithm

Let's do a brief thought experiment. What if we could rewrite the code that runs our news feeds and messaging apps from scratch, drawing from the enormous pool of content we generate every day? What if we could build an algorithm to maximize the protection of human choice? Let's examine some principles of what a great sensemaking algorithm could do if it was built for our flourishing instead of our outrage.

Some of its first principles might include the following:

- *Verification.* Ensuring all users are real humans, while protecting privacy and pseudonymity.
- *A clear, unambiguous constitution,* including an agreed-upon framework for updating content moderation policies and major site changes, which could be amended through due process by its users. It would provide meaningful opportunities for collective input on its ranking system.
- *Its ranking algorithm would be conditionally transparent.* It would consist of a framework available for study and examination through an

auditable API available to certain authorities. This would allow it to be regularly examined by teams of experts looking to ensure its own constitutional fidelity and monitor for bias and potential harms.

- *Integrity.* It would be structured against deceptive influence, and oriented toward reducing any risk of deliberate manipulation and system-wide gaming.
- *In the pursuit of truth, it could proportionately serve the most accurate information available.* It might parse complex threats and prioritize statistical facts above emotional anecdotes. It might also serve us proportional representations of the dangers facing us as individuals and communities. It would package news and issues in a way that provides deep context.
- *Moral translation.* It might encourage us to think through pressing issues with a dynamic moral lens, translating larger problems so they might appeal to our individual moral foundations.
- *Politics in good faith.* It could show the best version of opposing positions on controversial topics. It might work to facilitate consensus on hard but necessary moral actions by offering the best version of the opposing side's arguments on every contentious issue.
- *It could earn our attention with trust and openness*, not because it is the only tool available.

There is no precise blueprint that will solve all our current problems; this is merely an exercise in imagining what we might create. While some of these principles may seem at odds on their surface, thinking through the design of such a system is becoming a necessary task. We didn't get a chance to do this the first time around. Any platform that controls and influences humans at such an enormous scale must provide us with clear ways of understanding it, clear ways of managing it, and clear ways of ensuring that it serves our best interests.

Chapter 30

Coda

I grew up in a small town in northern California in the hills above wine country. My childhood home was a modest 1950s-era postwar structure surrounded by valley oaks and pine woodlands. In the depths of the pandemic in September of 2020, a fast-moving wildfire swept into my hometown and blew through the neighborhood I grew up in, destroying the house my parents had been living in for thirty-eight years.

My parents evacuated safely. Unfortunately all of my family's possessions were lost. Overnight, we watched decades of accumulated heirlooms, baby photos, journals, and memories disappear. It was a profound, traumatic blow, and one we are still processing years later. We lost the physical connection to our shared history.

In the wake of this sudden tragedy, something extraordinary happened on social media in the days after the fire: It showed us its best side. Beyond the expected condolences, a GoFundMe set up for my parents was widely shared across different messaging apps. An extended network of friends of friends reached out, offering anything they could to help their transition. The sudden and overwhelming outpouring of support—both emotional and financial— from friends and strangers alike turned a personal tragedy into a shared moment of feeling held by our community. A deep and tangible reflection of our community's bond with my family. Social media made this hard thing far more bearable. A moment of grace in the midst of a horrible time.

When these tools work, they can do incredible things. They become focal points for community support, and powerful places for catharsis and fellowship. In many ways, this was the original promise of social media— exactly where we began this book.

Throughout our journey, we've explored the many negative fractals of our media system and how it plays upon our outrage. Today's social media tools are, of course, not all bad. We use them because they provide us with real value, and show even greater promise. When I look to the future, I am still struck by their potential, and reminded of a passage by Emerson I love. It fills me with hope for what our media might become:

> We lie in the lap of immense intelligence, which makes us receivers of its truth and organ of its activity. When we discern justice, when we discern truth, we do nothing by ourselves, but allow a passage to its beams.

These words reference the emergent collective intelligence that might be felt when we are exposed to the natural world. When I get glimpses of the genuine potential of social media, I feel a similar sense of awe and wonder. I can see an image of what it might become—a tool to help us be better stewards of our shared existence. A tool for mutual comprehension. A tool that might help us see one another more honestly, and help us share responsibility for the world we mutually inhabit. I am still, after years of examining all of its faults, not so cynical. I know this is a tall order for a system with such a painful recent history, and we still have far to go. But I am hopeful we will find our way beyond this dark valley, because we have no choice but to climb out together.

Acknowledgments

This book would not have been possible without the support of many people, far too many to name here, and several that I certainly forgot.

Thank you to everyone who has inspired, encouraged, and believed in this project from the beginning.

My agent Jaidree Braddix and the Park & Fine team for being a great champion, and having my back through the most challenging moments.

The Nash family, for love and support, and for giving me solace and a quiet place to be with my creative self. Ellen Abrams and John Stossel, for great conversations and a lovely place to write.

Special thanks to my early readers, including Elyse DeBelser, Maria Bridge, Ilana Gilovich-Wave, Mark Fisher, Pippa Biddle, Julia Kamin, and Georgia Francis King for extracting a book from a relative mess.

My thought partners: Orion Henry, Max Stossel, Renée DiResta, Matteo Cantiello, Joshua Kauffman, Tristan Harris, Tim Urban, Chloé Valdary, Esther Perel, and Nicky Case, among others.

Jon Haidt, for encouraging me to dive into this journey headfirst, and for being a partner in deconstructing the machine in a way others might most easily understand.

My researchers: Alayna Kennedy, Anna Slavina. Special thanks to Zach Rausch for fantastic support during the last sprint to the finish.

My parents, for imbuing me with a love of writing and a critical eye for where I can improve. My mother, Judith, who gave weeks to the book, and whose ideas and thought partnership I cherish. My dad, Tom, who has always quietly been my biggest cheerleader.

My sister, for always being there.

My brothers Steve Martocci, Jonathan Swerdlin, Neil Parikh, Jared Matthew Weiss, Andrew Horn, David Yarus, Justin McLeod, Eli Clark-Davis, Matthew Kochmann, Adam Ward, Seth Miller, Chris Clement, and Ben Hindman for carrying me through some of the manuscript's hardest moments.

My amazing writing group: Chris Castiglione, Lexi Gervis, Quinn Simpson, Ryder Carroll, Ezzie Spencer, Quddus Philippe, Jade Tailor, Carmina Becerra, and Amber Rae.

My deeply supportive friends: Lindsay Ratowsky, Georgia Clark, Evan Walden, Mattan Griffel, Allie Hoffman, and the Agrawals for steadfast enthusiasm from the earliest days.

Thanks also to Krishan Trotman for believing so much in the project, Clarence, Amina, Carolyn, and the whole team at Hachette for their patience with this extremely complex topic and many, many edits.

My friends at Little Choc for smiles and espresso, which fueled many of my writing sessions.

Thank you to David Mindich, Michael Schudson Kevin Roose, Billy Brady, Logan Molyneaux, Erica Anderson, Bobby Bailey, and Ben Keesey, and so many others for generously agreeing to be interviewed and sharing your stories and expertise.

My friends in Cambodia, particularly Yinh Ya, Bryse Gaboury, Steve Forbes and the whole EWB team that helped build something truly improbable. My old school OK family, and the board and donors to the Human Translation Project.

And finally, Anneke Jong, for the love that has kept me upright.

Notes

Chapter 1: Empathy Machine

1 Elizabeth Becker, *When the War Was Over: Cambodia and the Khmer Rouge Revolution,* (New York: PublicAffairs, 1998), 39–41.

2 Ibid.

3 Steven Michael DeBurger, "The Khmer Rouge and the Re-Visioning of the Khmer Empire: Buddhism Encounters Political Religion," APSA 2011 Annual Meeting Paper, available at SSRN, https://papers.ssrn.com/abstract=1903043.

4 United States Holocaust Memorial Museum, "'Smashing' Internal Enemies," United States Holocaust Memorial Museum, April 2018, https://www.ushmm.org/genocide -prevention/countries/cambodia/case-study/violence/smashing-internal-enemies.

5 Nalini Vittal, "Tribulation before Trial in Cambodia: Confronting Autogenocide," *Economic and Political Weekly* 36, no. 3 (2001): 199–203, https://www.jstor.org/stable /4410192.

6 Becker, 1.

7 David Ashley, "Between War and Peace: Cambodia 1991–1998." In *Safeguarding Peace: Cambodia's Constitutional Challenge,* ed. Dylan Hendrickson (London: Conciliation Resources, 1998), Rich Garella and Eric Pape, "A Tragedy of No Importance," *Mother Jones* (April 15, 2005). https://www.motherjones.com/politics/2012/11 /cambodia-war-khmer-sam-rainsy/; Brendan Brady, "1994 Murder of Aussie by Khmer Rouge Re-Examined," *TIME* (March 2, 2010), https://content.time.com/time/world /article/0,8599,1968996,00.html.

8 Aly Weisman, "CHART: How #Kony2012 Just Became the Most Viral Video of All Time," *Business Insider* (March 12, 2012). https://www.businessinsider.com/how-kony2012-just -became-the-most-viral-video-of-all-time-2012-3; Emma Madden, "'Kony 2012,' 10 Years Later," *New York Times* (March 8, 2022). https://www.nytimes.com/2022/03/08/style/kony -2012-invisible-children.html.

9 Barack Obama, "Obama to Graduates: Cultivate Empathy: Northwestern University News," *Northwestern News* (June 19, 2006). https://www.northwestern.edu/newscenter /stories/2006/06/barack.html.

10 Jose Antonio Vargas, "Spring Awakening," *New York Times* (February 17, 2012). https:// www.nytimes.com/2012/02/19/books/review/how-an-egyptian-revolution-began-on -facebook.html.

11   Spencer Kornhaber, "Is Empathy Overrated?" *Atlantic* (July 3, 2015). https://www
.theatlantic.com/health/archive/2015/07/against-empathy-aspen-paul-bloom-richard
-j-davidson/397694/.

12   Paul Bloom, *Against Empathy: The Case for Rational Compassion* (New York: Ecco,
2016), 152.

13   WAN-IFRA Staff, "Upworthy's Most Successful Strategy Ever." World Association of
New Publishers News (December 6, 2013). https://wan-ifra.org/2013/12/upworthys-most
-successful-strategy-ever/; Jeff Bercovici, "These Five Astonishing Headline Writing
Secrets Will Make You Cry, Or At Least Click," *Forbes* (March 1, 2013). https://www
.forbes.com/sites/jeffbercovici/2013/03/01/these-five-astonishing-headline-writing
-secrets-will-make-you-cry/; Anya Kamenetz, "How Upworthy Used Emotional Data to
Become the Fastest Growing Media Site of All Time," *Fast Company* (June 7, 2013). https://
www.fastcompany.com/3012649/how-upworthy-used-emotional-data-to-become
-the-fastest-growing-media-site-of-all-time.

Chapter 2: The Feed

1   Andrew Pettegree, *The Invention of News: How the World Came to Know about Itself*
(New Haven, CT: Yale University Press, 2014).

Chapter 3: The Overwhelming Present

1   Mihaly Csikszentmihalyi, "Flow: The Psychology of Optimal Experience," in *Flow: The
Psychology of Optimal Experience*, Nachdr., Harper Perennial Modern Classics (New
York: Harper Collins, 2009), 28–29.

2   Daniel J. Levitin, *The Organized Mind: Thinking Straight in the Age of Information Over-
load* (New York: Dutton, 2014).

3   Levitin, 41.

4   Jeffrey T. Klein, Stephen V. Shepherd, and Michael L. Platt, "Social Attention and
the Brain," *Current Biology* 19, no. 20 (November 3, 2009): R958–62, https://doi
.org/10.1016/j.cub.2009.08.010.

5   "History of the Web," World Wide Web Foundation. https://webfoundation.org/about
/vision/history-of-the-web/.

6   Ethan Zuckerman, "The Internet's Original Sin," *The Atlantic*, August 14, 2014, https://www
.theatlantic.com/technology/archive/2014/08/advertising-is-the-internets-original-sin
/376041/.

7   Sean Parker, "The Epic Tale of MySpace's Technical Failure," Wiredelta, May 17, 2019,
https://wiredelta.com/the-epic-tale-myspace-technical-failure/; "MySpace and the Coding
Legacy It Left Behind," *Codecademy* (blog), February 14, 2020, https://www.codecademy
.com/resources/blog/myspace-and-the-coding-legacy/.

8   Tobias Rose-Stockwell and Jonathan Haidt, "The Dark Psychology of Social Networks," *The
Atlantic*, November 12, 2019, https://www.theatlantic.com/magazine/archive/2019/12
/social-media-democracy/600763/.

9   Daniel Kahneman, *Thinking, Fast and Slow* (London: Penguin Books, 2012).

10   Renée Diresta and Tobias Rose-Stockwell, "How to Stop Misinformation Before It Gets Shared," *Wired*, March 26, 2021, https://www.wired.com/story/how-to-stop-misinformation-before-it-gets-shared/.

11   Soroush Vosoughi, Deb Roy, and Sinan Aral, "The Spread of True and False News Online," *Science* 359, no. 6380 (March 9, 2018): 1146–51, https://doi.org/10.1126/science.aap9559. Fake news's viral advantage is particularly true for certain types of misinformation. One study found that older users and conservatives are more likely to share fake news, particularly if the news is ideologically congenial. For older Americans, political misinformation is more popular whereas younger users are more likely to amplify clickbait. See Andy Guess et al., "Cracking Open the News Feed: Exploring What U.S. Facebook Users See and Share with Large-Scale Platform Data," *Journal of Quantitative Description: Digital Media* 1 (April 2021), https://doi.org/10.51685/jqd.2021.006.

12   Daniel Engber, "Sorry, I Lied About Fake News," *The Atlantic*, March 26, 2022, https://www.theatlantic.com/technology/archive/2022/03/fake-news-misinformation-mit-study/629396/.

Chapter 4: The Origin of Our Addictions

1   Steven Pinker, "Correct for the Media's Negativity Bias," *Politico*, 2019, https://politico.com/interactives/2019/how-to-fix-politics-in-america/misinformation/correct-for-the-medias-negativity-bias/.

2   Gary W. Small et al., "Brain Health Consequences of Digital Technology Use," *Dialogues in Clinical Neuroscience* 22, no. 2 (June 30, 2020): 179–87, https://doi.org/10.31887/DCNS.2020.22.2/gsmall.

3   Ryder Carroll, "How ADHD Helped Me Create the Bullet Journal Method," *Human Parts* (blog), November 25, 2019, https://humanparts.medium.com/inside-adhd-55b96 18cd708.

4   Brian A. Primack et al., "Social Media Use and Perceived Social Isolation among Young Adults in the U.S.," *American Journal of Preventive Medicine* 53, no. 1 (July 1, 2017): 1–8, https://doi.org/10.1016/j.amepre.2017.01.010; Melissa G. Hunt et al., "No More FOMO: Limiting Social Media Decreases Loneliness and Depression," *Journal of Social and Clinical Psychology* 37, no. 10 (December 2018): 751–68, https://doi.org/10.1521/jscp.2018.37.10.751; "2018 Global Mobile Consumer Survey: US Edition" (Deloitte, 2018), https://www2.deloitte.com/content/dam/Deloitte/us/Documents/technology-media-telecommunications/us-tmt-global-mobile-consumer-survey-exec-summary-2018.pdf; Lauren Hale et al., "Media Use and Sleep in Teenagers: What Do We Know?" *Current Sleep Medicine Reports* 5, no. 3 (September 1, 2019): 128–34, https://doi.org/10.1007/s40675-019-00146-x.

5   Tristan Harris, "Smartphone Addiction Is Part of the Design," *Der Spiegel*, July 27, 2016, sec. International, https://www.spiegel.de/international/zeitgeist/smartphone-addiction-is-part-of-the-design-a-1104237.html.

6   Ezra Klein, "Is Big Tech Addictive? A Debate with Nir Eyal," *Vox*, August 7, 2019, https://www.vox.com/podcasts/2019/8/7/20750214/nir-eyal-tech-addiction-ezra-klein-smartphones-hooked-indistractable.

Chapter 5: Pushing the Trigger

1 Jeremy Littau, "The Crisis Facing American Journalism Did Not Start with the Internet," *Slate*, January 26, 2019, https://slate.com/technology/2019/01/layoffs-at-media -organizations-the-roots-of-this-crisis-go-back-decades.html.

2 "How to Make That One Thing Go Viral," *Upworthy*, December 3, 2012, https://www .slideshare.net/Upworthy/how-to-make-that-one-thing-go-viral-just-kidding.

3 Hayley Tsukayama, "Facebook Reaches 1 Billion Users," *Washington Post*, October 4, 2012, https://www.washingtonpost.com/business/technology/facebook-reaches-1-billion -users/2012/10/04/5edfefb2-0e14-11e2-bb5e-492c0d30bff6_story.html.

4 Stefan Feuerriegel et al., "Negativity Drives Online News Consumption [Registered Report Stage 1 Protocol]," April 26, 2022, https://doi.org/10.6084/m9.figshare.19657452.v1.

5 Felix Salmon, "Headlines Matter," *Nieman Lab* (blog), accessed January 9, 2023, https://www.niemanlab.org/2016/12/headlines-matter/.

6 Louise Linehan, Steve Rayson, and Henley Wing Chiu, "100m Articles Analyzed: What You Need to Write the Best Headlines [2021]," BuzzSumo.com, August 17, 2021, https://buzzsumo.com/blog/most-shared-headlines-study/.

7 Ullrich K. H. Ecker et al., "The Effects of Subtle Misinformation in News Headlines," *Journal of Experimental Psychology: Applied* 20 (2014): 323–35, https://doi.org/10.1037 /xap0000028.

8 New York–based media executive, personal correspondence, February 22, 2017.

9 Paul Farhi, "One Billion Dollars Profit? Yes, the Campaign Has Been a Gusher for CNN," *Washington Post*, October 27, 2016, sec. Style, https://www.washingtonpost .com/lifestyle/style/one-billion-dollars-profit-yes-the-campaign-has-been-a-gusher-for -cnn/2016/10/27/1fc879e6-9c6f-11e6-9980-50913d68eacb_story.html.

10 Howard Fineman, "Look Who's Running," *Newsweek*, October 10, 1999, https://www .newsweek.com/look-whos-running-168162.

11 Nicholas Confessore and Karen Yourish, "$2 Billion Worth of Free Media for Donald Trump," *New York Times*, March 15, 2016, sec. The Upshot, https://www.nytimes.com/2016 /03/16/upshot/measuring-donald-trumps-mammoth-advantage-in-free-media.html.

12 Tobias Rose-Stockwell, "This Is How Your Fear and Outrage Are Being Sold for Profit," *Quartz*, July 28, 2017, https://qz.com/1039910/how-facebooks-news-feed-algorithm-sells -our-fear-and-outrage-for-profit/.

13 Kevin Roose, *New York Times*, personal correspondence, April 11, 2019.

14 Shannon C. McGregor and Logan Molyneux, "Twitter's Influence on News Judgment: An Experiment among Journalists," *Journalism* 21, no. 5 (May 1, 2020): 597–613, https://doi.org/10.1177/1464884918802975.

15 Madison Hall, "USA Today Wrapped Its Newspaper with a Fake Cover about 'Hybrid Babies' with Antlers to Advertise a New Netflix Show," *Insider*, accessed January 10, 2023, https://www.insider.com/usa-today-fake-cover-hybrid-babies-netflix-show-2021-6.

16 Anemona Hartocollis, "Craig Spencer, New York Doctor with Ebola, Will Leave Bellevue Hospital," CNBC, November 10, 2014, https://www.cnbc.com/2014/11/10/craig -spencer-new-york-doctor-with-ebola-will-leave-bellevue-hospital.html.

17　Ashley Collman, "Ebola Fears in America Reach New Levels as Woman in Hazmat Suit Waits for Plane at Washington Dulles," *Daily Mail*, October 16, 2014, https://www.dailymail.co.uk/news/article-2794947/not-taking-risks-woman-hazmat-suit-waits-plane-washington-dulles-airport-two-days-second-nurse-test-positive-ebola-boarded-flight-fever.html; Alice Ritchie, "Ebola Is 'Disaster of Our Generation' Says Aid Agency," Yahoo News, October 18, 2014, http://news.yahoo.com/obama-calls-end-ebola-hysteria-110006011.html.

18　John Gramlich, "Violent Crime Is a Key Midterm Voting Issue, but What Does the Data Say?" Pew Research Center, October 31, 2022, https://www.pewresearch.org/fact-tank/2022/10/31/violent-crime-is-a-key-midterm-voting-issue-but-what-does-the-data-say/.

19　Ibid. The US murder rate in 2020 was 42 percent lower than the suicide rate (13.5 deaths per 100,000 people) and 71 percent below the mortality rate for drug overdose (27.1 deaths per 100,000 people, as of the third quarter of 2020), the CDC data shows.

20　Alberto M. Fernandez, "Here to Stay and Growing: Combating ISIS Propaganda Networks," Project on U.S. Relations with the Islamic World, Center for Middle East Policy at Brookings (Center for Middle East Politics, Brookings Institution, October 2015), https://www.brookings.edu/wp-content/uploads/2016/07/IS-Propaganda_Web_English_v2.pdf.

21　Office of Public Affairs, "Man Sentenced to Life in Prison for ISIS-Inspired Bombing in New York City Subway Station in 2017," United States Department of Justice, April 22, 2021, https://www.justice.gov/opa/pr/man-sentenced-life-prison-isis-inspired-bombing-new-york-city-subway-station-2017.

22　Daniel T. Blumstein, *The Nature of Fear: Survival Lessons from the Wild* (Cambridge, MA: Harvard University Press, 2020).

23　Amos Tversky and Daniel Kahneman, "Availability: A Heuristic for Judging Frequency and Probability," *Cognitive Psychology* 5, no. 2 (September 1, 1973): 207–32, https://doi.org/10.1016/0010-0285(73)90033-9.

24　"Malaria Worldwide—Impact of Malaria," Centers for Disease Control and Prevention, December 16, 2021, https://www.cdc.gov/malaria/malaria_worldwide/impact.html.

25　Penelope Muse Abernathy, "The Expanding News Desert" (Center for Innovation and Sustainability in Local Media, University of North Carolina at Chapel Hill, 2018), https://www.usnewsdeserts.com/reports/expanding-news-desert/; Neal Rothschild and Sara Fischer, "News Engagement Plummets as Americans Tune Out," *Axios*, July 12, 2022, https://www.axios.com/2022/07/12/news-media-readership-ratings-2022.

Chapter 6: Black and Blue, White and Gold

1　Claudia Koerner, "The Dress Is Blue and Black, Says the Girl Who Saw It in Person," *BuzzFeed News*, February 26, 2015, https://www.buzzfeednews.com/article/claudiakoerner/the-dress-is-blue-and-black-says-the-girl-who-saw-it-in-pers.

2　Cates Holderness, "What Colors Are This Dress?" *BuzzFeed*, February 26, 2015, https://www.buzzfeed.com/catesish/help-am-i-going-insane-its-definitely-blue.

3   Rosa Lafer-Sousa, Katherine L. Hermann, and Bevil R. Conway, "Striking Individual Differences in Color Perception Uncovered by 'the Dress' Photograph," *Current Biology* 25, no. 13 (June 2015): R545–46, https://doi.org/10.1016/j.cub.2015.04.053; Pascal Wallisch, "Illumination Assumptions Account for Individual Differences in the Perceptual Interpretation of a Profoundly Ambiguous Stimulus in the Color Domain: 'The Dress,'" *Journal of Vision* 17, no. 4 (June 12, 2017): 5, https://doi.org/10.1167/17.4.5.

4   Scott Alexander, "Sort by Controversial," *Slate Star Codex* (blog), October 31, 2018, https://slatestarcodex.com/2018/10/30/sort-by-controversial/.

5   Richard Dawkins, *The Selfish Gene: 40th Anniversary Edition* (New York: Oxford University Press, 2016).

6   Susan Blackmore, *The Meme Machine* (Oxford, UK: Oxford University Press, 2000).

Chapter 7: The Engagement Escalator

1   Mark Zuckerberg, "A Blueprint for Content Governance and Enforcement," Meta, May 5, 2021, https://www.facebook.com/notes/751449002072082/.

2   Adam Kramer, "The Spread of Emotion via Facebook," *Meta Research*, May 16, 2012, https://research.facebook.com/publications/the-spread-of-emotion-via-facebook/.

3   William J. Brady, M. J. Crockett, and Jay J. Van Bavel, "The MAD Model of Moral Contagion: The Role of Motivation, Attention, and Design in the Spread of Moralized Content Online," *Perspectives on Psychological Science* 15, no. 4 (July 1, 2020): 978–1010, https://doi.org/10.1177/1745691620917336.

4   Frans B. M. de Waal, "The Antiquity of Empathy," *Science* 336, no. 6083 (May 18, 2012): 874–76, https://doi.org/10.1126/science.1220999.

5   Wataru Nakahashi and Hisashi Ohtsuki, "Evolution of Emotional Contagion in Group-Living Animals," *Journal of Theoretical Biology* 440 (March 7, 2018): 12–20, https://doi.org/10.1016/j.jtbi.2017.12.015.

6   Adam D. I. Kramer, Jamie E. Guillory, and Jeffrey T. Hancock, "Experimental Evidence of Massive-Scale Emotional Contagion through Social Networks," *Proceedings of the National Academy of Sciences* 111, no. 24 (June 17, 2014): 8788–90, https://doi.org/10.1073/pnas.1320040111.

7   Joseph G. Lehman, "An Introduction to the Overton Window of Political Possibility," Mackinac Center, April 8, 2010, https://www.mackinac.org/12481. After Overton's death in 2003, Joseph G. Lehman and others developed the term "Overton Window" to describe their late colleague's theory of change.

8   Derek Robertson, "How an Obscure Conservative Theory Became the Trump Era's Go-to Nerd Phrase," *Politico Magazine*, February 25, 2018, https://www.politico.com/magazine/story/2018/02/25/overton-window-explained-definition-meaning-217010/.

9   "Watch Live: Facebook CEO Zuckerberg Speaks at Georgetown University," streamed live by *the Washington Post* on October 17, 2019, https://www.youtube.com/watch?v=2MTpd7YOnyU.

Chapter 8: The Apple of Discord

1   Charity Davenport, "Story: The Trojan War Part 1: The Apple of Discord," September 8, 2018, https://pressbooks.pub/iagtm/chapter/story-the-trojan-war/.
2   Emily Stewart, "Covington Catholic Students' Clash with a Native American Elder, Explained," *Vox*, January 24, 2019, https://www.vox.com/2019/1/22/18192908/covington-catholic-video-nick-sandmann-maga.
3   "Video Shows Different Side of Controversial Viral Video," *CNN Newsroom*, accessed January 10, 2023, https://www.cnn.com/videos/us/2019/01/23/maga-teens-covington-catholic-native-american-sidner-pkg-nr.cnn.
4   Craig Silverman, "Lies, Damn Lies, and Viral Content," *Columbia Journalism Review*, February 10, 2015, https://www.cjr.org/tow_center_reports/craig_silverman_lies_damn_lies_viral_content.php/.

Chapter 9: Trigger-Chain

1   William J. Brady et al., "Emotion Shapes the Diffusion of Moralized Content in Social Networks," *Proceedings of the National Academy of Sciences* 114, no. 28 (July 11, 2017): 7313–18, https://doi.org/10.1073/pnas.1618923114.
2   Michael Macy et al., "Opinion Cascades and the Unpredictability of Partisan Polarization," *Science Advances* 5, no. 8 (August 28, 2019): eaax0754, https://doi.org/10.1126/sciadv.aax0754.
3   Steven Pinker, *The Blank Slate: The Modern Denial of Human Nature*, Nachdr. (London: Penguin, 2003), 286.
4   Macy et al., 2019.

Chapter 10: Algorithms

1   Adam Mosseri, "Bringing People Closer Together," *Meta* (blog), January 12, 2018, https://about.fb.com/news/2018/01/news-feed-fyi-bringing-people-closer-together/.
2   Shoshana Zuboff, *The Age of Surveillance Capitalism* (New York: PublicAffairs, 2017), 75.
3   Eric Siegel, "When Does Predictive Technology Become Unethical?" *Harvard Business Review*, October 23, 2020, https://hbr.org/2020/10/when-does-predictive-technology-become-unethical.
4   R. F. C. Hull, ed., "II The Shadow," in *Collected Works of C.G. Jung, Volume 9 (Part 2): Aion: Researches into the Phenomenology of the Self*, by C. G. Jung (Princeton, NJ: Princeton University Press, 2014), 8–10, https://doi.org/10.1515/9781400851058.8.
5   Robert Epstein and Ronald E. Robertson, "The Search Engine Manipulation Effect (SEME) and Its Possible Impact on the Outcomes of Elections," *Proceedings of the National Academy of Sciences* 112, no. 33 (August 18, 2015): E4512–21, https://doi.org/10.1073/pnas.1419828112.
6   Olivia Solon and Sabrina Siddiqui, "Russia-Backed Facebook Posts 'Reached 126m Americans' during US Election," *Guardian*, October 31, 2017, sec. Technology, https://www.theguardian.com/technology/2017/oct/30/facebook-russia-fake-accounts-126

-million; "Update on Twitter's Review of the 2016 US Election," *Twitter Blog*, January 19, 2019, https://blog.twitter.com/en_us/topics/company/2018/2016-election-update.

7   Yuval Noah Harari, *Homo Deus* (London: Harvill Secker, 2016).

8   Samuel Gibbs, "AlphaZero AI Beats Champion Chess Program after Teaching Itself in Four Hours," *Guardian*, December 7, 2017, sec. Technology, https://www.theguardian.com/technology/2017/dec/07/alphazero-google-deepmind-ai-beats-champion-program-teaching-itself-to-play-four-hours.

Chapter 11: Intuitions and the Internet

1   Rachel O'Donoghue, "Pit Bulls Pull Owner to Ground as They Brutally Maul Cat in Horrific Attack," *Mirror*, July 7, 2017, sec. UK News, http://www.mirror.co.uk/news/uk-news/vicious-pit-bulls-pull-helpless-10755320.

2   Niall McCarthy, "Infographic: America's Most Dangerous Dog Breeds," Statista Infographics, September 14, 2018, https://www.statista.com/chart/15446/breeds-of-dog-involved-in-fatal-attacks-on-humans-in-the-us.

3   R. A. Casey et al., "Inter-Dog Aggression in a UK Owner Survey: Prevalence, Co-Occurrence in Different Contexts and Risk Factors," *Veterinary Record* 172, no. 5 (2013): 127–127, https://doi.org/10.1136/vr.100997.

4   American Veterinary Medical Association Animal Welfare Division, "Dog Bite Risk and Prevention: The Role of Breed," May 15, 2014, https://www.avma.org/resources-tools/literature-reviews/dog-bite-risk-and-prevention-role-breed.

5   R. B. Zajonc, "Feeling and Thinking: Preferences Need No Inferences," *American Psychologist* 35 (1980): 151–75, https://doi.org/10.1037/0003-066X.35.2.151.

6   Jonathan Haidt, "The Emotional Dog and Its Rational Tail: A Social Intuitionist Approach to Moral Judgment," *Psychological Review* 108 (2001): 814–34, https://doi.org/10.1037/0033-295X.108.4.814.

7   Jennifer S. Lerner et al., "Emotion and Decision Making," *Annual Review of Psychology* 66, no. 1 (2015): 799–823, https://doi.org/10.1146/annurev-psych-010213-115043.

8   Jonathan Haidt, "The Moral Emotions," in *Handbook of Affective Sciences*, Series in Affective Science (New York: Oxford University Press, 2003), 852–70.

9   Jesse Graham et al., "Chapter Two—Moral Foundations Theory: The Pragmatic Validity of Moral Pluralism," in *Advances in Experimental Social Psychology*, ed. Patricia Devine and Ashby Plant, vol. 47 (Cambridge, MA: Academic Press, 2013), 55–130, https://doi.org/10.1016/B978-0-12-407236-7.00002-4.

10 Jonathan Haidt, ed., *The Righteous Mind: Why Good People Are Divided by Politics and Religion*, (New York: Vintage Books, 2013), 21–60.

11 Ibid., 16.

12 Ibid., 125.

13 Gary F. Marcus, *The Birth of the Mind: How a Tiny Number of Genes Creates the Complexities of Human Thought* (New York: Basic Books, 2004), 40.

14 Elias Dinas, "Why Does the Apple Fall Far from the Tree? How Early Political Socialization Prompts Parent-Child Dissimilarity," *British Journal of Political Science* 44, no. 4 (October 2014): 827–52, https://doi.org/10.1017/S0007123413000033.

15 John C. Turner et al., *Rediscovering the Social Group: A Self-Categorization Theory*, (Cambridge, MA: Basil Blackwell, 1987); Michael A. Hogg, "Social Identity Theory," in *Understanding Peace and Conflict Through Social Identity Theory: Contemporary Global Perspectives*, ed. Shelley McKeown, Reeshma Haji, and Neil Ferguson, Peace Psychology Book Series (Cham, Switzerland: Springer International Publishing, 2016), 3–17, https://doi.org/10.1007/978-3-319-29869-6_1.

16 Maeve Duggan and Aaron Smith, "The Political Environment on Social Media" (Pew Research Center, October 25, 2016), https://www.pewresearch.org/internet/2016/10/25/the-political-environment-on-social-media/.

17 Jonathan Haidt, "Why the Past 10 Years of American Life Have Been Uniquely Stupid," *The Atlantic*, April 11, 2022, https://www.theatlantic.com/magazine/archive/2022/05/social-media-democracy-trust-babel/629369/.

Chapter 12: The Worst Room

1 Shay Maunz, "The Great Hanoi Rat Massacre of 1902 Did Not Go as Planned," *Atlas Obscura*, June 6, 2017, http://www.atlasobscura.com/articles/hanoi-rat-massacre-1902.

2 Stanley Milgram, Leonard Bickman, and Lawrence Berkowitz, "Note on the Drawing Power of Crowds of Different Size," *Journal of Personality and Social Psychology* 13 (1969): 79–82, https://doi.org/10.1037/h0028070.

3 Thomas McMullan, "The Inventor of the Facebook Like: 'There's Always Going to Be Unintended Consequences,'" *Alphr* (blog), October 20, 2017, https://www.alphr.com/facebook/1007431/the-inventor-of-the-facebook-like-theres-always-going-to-be-unintended-consequences/.

4 B. F. Skinner, "'Superstition' in the Pigeon," *Journal of Experimental Psychology* 38 (1948): 168–72, https://doi.org/10.1037/h0055873.

5 Robert D. Pritchard et al., "The Effects of Varying Schedules of Reinforcement on Human Task Performance," *Organizational Behavior and Human Performance* 16, no. 2 (August 1, 1976): 205–30, https://doi.org/10.1016/0030-5073(76)90014-3; J. E. R. Staddon and D. T. Cerutti, "Operant Conditioning," *Annual Review of Psychology* 54 (2003): 115–44, https://doi.org/10.1146/annurev.psych.54.101601.145124.

6 William J. Brady et al., "How Social Learning Amplifies Moral Outrage Expression in Online Social Networks," *Science Advances* 7, no. 33 (August 13, 2021): eabe5641, https://doi.org/10.1126/sciadv.abe5641.

7 Bill Hathaway, "'Likes' and 'Shares' Teach People to Express More Outrage Online," *YaleNews*, August 13, 2021, https://news.yale.edu/2021/08/13/likes-and-shares-teach-people-express-more-outrage-online.

8 "The Influencer Report: Engaging Gen Z and Millenials" (Morning Consult, 2022), https://morningconsult.com/influencer-report-engaging-gen-z-and-millennials/.

9 Mark R. Leary, "Sociometer Theory and the Pursuit of Relational Value: Getting to the Root of Self-Esteem," *European Review of Social Psychology* 16, no. 1 (January 1, 2005): 75–111, https://doi.org/10.1080/10463280540000007.

10 Justin Tosi and Brandon Warmke, *Grandstanding: The Use and Abuse of Moral Talk* (New York: Oxford University Press, 2020).

11 Paul Graham, "How to Disagree," March 2008, http://www.paulgraham.com/disagree.html.

12 C. Thi Nguyen, "How Twitter Gamifies Communication," in *Applied Epistemology*, ed. Jennifer Lackey (Oxford, UK: Oxford University Press, 2021), 410–36, https://doi.org/10.1093/oso/9780198833659.003.0017.

13 Johan Huizinga, *Homo Ludens: A Study of the Play-Element in Culture* (London: Routledge and Kegan Paul, 1949).

14 Shopify Staff, "Influencer Marketing Prices: How Much Should You Pay (2023)," *Shopify Blog*, December 5, 2022, https://www.shopify.com/blog/influencer-pricing.

15 Statista Research Team, "Global Twitter CPM 2020," Statista, June 2020, https://www.statista.com/statistics/872543/twitter-cost-per-mile/.

16 William J. Brady et al., "Emotion Shapes the Diffusion of Moralized Content in Social Networks," *Proceedings of the National Academy of Sciences* 114, no. 28 (July 11, 2017): 7313–18, https://doi.org/10.1073/pnas.1618923114.

17 Tobias Rose-Stockwell, "How We Broke Democracy (But Not in the Way You Think)," *Medium* (blog), February 11, 2019, https://tobiasrose.medium.com/empathy-to-democracy-b7f04ab57eee.

18 Steven Asarch et al., "Inside the Rise of Nikocado Avocado, the Extreme-Eating YouTuber Whose Dramatic Meltdowns Have Led to Years of Controversy and Feuds," *Insider*, November 25, 2022, https://www.insider.com/who-is-youtube-star-nikocado-avocado-2020-1.

19 Boghal, "The Perils of Audience Capture," Substack newsletter, *The Prism* (blog), June 30, 2022, https://gurwinder.substack.com/p/the-perils-of-audience-capture.

20 Ibid.

Chapter 13: Trauma, Processing, and Cancellation

1 Lee Moran, "Stephen Colbert under Fire for Comedy Central's 'Racist' Tweet Based on Satirical Asian Skit," *New York Daily News*, March 28, 2014, https://www.nydailynews.com/entertainment/tv-movies/stephen-colbert-fire-comedy-central-racist-tweet-article-1.1737621.

2 Alyssa Rosenberg, "Stephen Colbert Was Making Fun of Dan Snyder, Not Asians and Asian-Americans," *Washington Post*, March 28, 2014, https://www.washingtonpost.com/news/act-four/wp/2014/03/28/stephen-colbert-was-making-fun-of-dan-snyder-not-asians-and-asian-americans/; Jay Caspian Kang, "The Campaign to 'Cancel' Colbert," *New Yorker*, March 30, 2014, https://www.newyorker.com/news/news-desk/the-campaign-to-cancel-colbert.

3 *The Internet Ruined My Life*, SYFY, March 2016.

4 *Hannah Gadsby: Nanette*, 2018, https://www.netflix.com/title/80233611.

5 The full APA definition of "trauma" is, "an emotional response to a terrible event like an accident, rape, or natural disaster. Immediately after the event, shock and denial are typical. Longer-term reactions include unpredictable emotions, flashbacks, strained relationships, and even physical symptoms like headaches or nausea." Gerasimos Kolaitis and Miranda Olff, "Psychotraumatology in Greece," *European*

Journal of Psychotraumatology 8, no. sup4 (September 29, 2017): 135175, https://doi .org/10.1080/20008198.2017.1351757; "Trauma," American Psychological Association, accessed January 10, 2023, https://www.apa.org/topics/trauma.

6   Holly Muir and Spencer Greenberg, "Understanding Relationship Conflicts: Clashing Trauma," Clearer Thinking, May 5, 2022, https://www.clearerthinking.org/post /understanding-relationship-conflicts-clashing-trauma.

7   Tara Brach, "The Trance of Fear," *Tara Brach* (blog), July 25, 2013, https://www .tarabrach.com/the-trance-of-fear/.

8   Teah Strozer, "Life Hurts: Responding with RAIN with Teah Strozer," *Tricycle: The Buddhist Review*, April 2015, https://tricycle.org/dharmatalks/life-hurts-responding-rain/.

9   Gilbert Gottfried, "Gilbert Gottfried on His Infamous 9/11 Joke and 'Too Soon,' " *Vulture*, February 2, 2016, https://www.vulture.com/2016/02/gilbert-gottfried-on-his -911-joke-too-soon.html.

10 Roy F. Baumeister, Liqing Zhang, and Kathleen D. Vohs, "Gossip as Cultural Learning," *Review of General Psychology* 8, no. 2 (June 1, 2004): 111–21, https://doi .org/10.1037/1089-2680.8.2.111; Francesca Giardini et al., "Gossip and Competitive Altruism Support Cooperation in a Public Good Game," *Philosophical Transactions of the Royal Society B: Biological Sciences* 376, no. 1838 (October 4, 2021): 20200303, https://doi.org/10.1098/rstb.2020.0303.

11 Foundation for Individual Rights and Expression, "Campus Disinvitation Database," accessed January 10, 2023, https://www.thefire.org/research-learn/campus-disinvitation -database.

12 Émile Durkheim and Steven Lukes, *The Division of Labour in Society*, trans. W. D. Halls, 2nd ed. (Basingstoke, UK: Palgrave Macmillan, 2013).

13 FIRE, "Cancel Culture Widely Viewed as Threat to Democracy, Freedom," Foundation for Individual Rights and Expression, *FIRE Newsdesk* (blog), January 31, 2022, https://www.thefire.org/news/national-fire-survey-cancel-culture-widely-viewed -threat-democracy-freedom.

14 "2020 College Free Speech Rankings," Foundation for Individual Rights and Expression, 2020, https://www.thefire.org/research-learn/2020-college-free-speech-rankings.

Chapter 14: The Waves of Moral Norms

1   Georg Wilhelm Friedrich Hegel et al., *The Encyclopaedia Logic, with the Zusätze: Part I of the Encyclopaedia of Philosophical Sciences with the Zusätze* (Indianapolis: Hackett, 1991).

2   "Yearly Number of Animals Slaughtered for Meat," Our World in Data, accessed January 10, 2023, https://ourworldindata.org/grapher/animals-slaughtered-for-meat.

3   Ralph Waldo Emerson, *The Conduct of Life* (Boston: Ticknor and Fields, 1863).

4   Anugraha Sundaravelu, "Earth's Ozone Layer Continues Heal in 2022, Says NASA" *METRO News*, October 28, 2022, https://metro.co.uk/2022/10/28/earths -ozone-layer-continues-heal-in-2022-says-nasa-17656122/.

5   Hunter Oatman-Stanford, "What Were We Thinking? The Top 10 Most Dangerous Ads," *Collectors Weekly*, August 22, 2012, https://www.collectorsweekly.com/articles /the-top-10-most-dangerous-ads/; "An Ad for Iver Johnson Revolvers From 1904

Claimed to Be Safe Enough to Be Near Babies," December 2020, https://www.vintag.es/2020/09/1904-iver-johnson-revolver-ad.html.

6   Alexander Hamilton, James Madison, and John Jay, eds., "No. 1: General Introduction," in *The Federalist Papers*, Dover Thrift Editions (Mineola, New York: Dover Publications, Inc, 2014), 5.

7   Nick Thompson, "Benedict Cumberbatch Sorry for 'Colored Actors' Remark," *CNN Entertainment*, January 28, 2015, https://www.cnn.com/2015/01/27/entertainment/benedict-cumberbatch-colored-apology/index.html.

Chapter 15: The Dark Valley

1   Socheata Sann et al., "Sociological Analysis of the Road Safety Situation in Cambodia: Historical, Cultural, and Political Aspects," no. 79 (2009).

2   Matthew Sparkes, "Bitcoin Has Emitted 200 Million Tonnes of CO2 Since Its Launch," *New Scientist*, September 28, 2022, https://www.newscientist.com/article/2339629-bitcoin-has-emitted-200-million-tonnes-of-co2-since-its-launch/.

3   J. Arjan G. M. de Visser et al., "The Utility of Fitness Landscapes and Big Data for Predicting Evolution," *Heredity* 121, no. 5 (November 2018): 401–5, https://doi.org/10.1038/s41437-018-0128-4.

4   Deborah Blum, "Looney Gas and Lead Poisoning: A Short, Sad History," *Wired*, January 5, 2013, https://www.wired.com/2013/01/looney-gas-and-lead-poisoning-a-short-sad-history/.

5   Nicholas Rees and Richard Fuller, "The Toxic Truth: Children's Exposure to Lead Pollution Undermines a Generation of Future Potential" (Unicef, 2020), https://www.unicef.org/sites/default/files/2020-07/The-toxic-truth-children%E2%80%99s-exposure-to-lead-pollution-2020.pdf.

6   Blum, 2013.

7   Kelsey Piper, "One of the Worst Public Health Dangers of the Past Century Has Finally Been Eradicated," *Vox*, September 3, 2021, https://www.vox.com/future-perfect/22650920/leaded-gasoline-eradicated-public-health.

8   "How Safe Is Nuclear Energy?" *Economist*, July 19, 2022, https://www.economist.com/graphic-detail/2022/07/19/how-safe-is-nuclear-energy.

9   Sarah Kramer, "Here's Why a Chernobyl-Style Nuclear Meltdown Can't Happen in the United States," *Business Insider*, April 26, 2016, https://www.businessinsider.com/chernobyl-meltdown-no-graphite-us-nuclear-reactors-2016-4; Hannah Ritchie, "What Was the Death Toll from Chernobyl and Fukushima?" Our World in Data, July 24, 2017, https://ourworldindata.org/what-was-the-death-toll-from-chernobyl-and-fukushima.

10   Eli Pariser, *The Filter Bubble: How the New Personalized Web Is Changing What We Read and How We Think*, 2nd printing (London: Penguin Books, 2012).

11   Gordon W. Allport, *The Nature of Prejudice*, Unabridged, 25th anniversary ed. (Reading, MA: Addison-Wesley, 1979).

12   Christopher A. Bail et al., "Exposure to Opposing Views on Social Media Can Increase Political Polarization," *Proceedings of the National Academy of Sciences* 115, no. 37 (September 11, 2018): 9216–21, https://doi.org/10.1073/pnas.1804840115.

Chapter 16: The Ancient History of Virality

1   "News | Etymology, Origin and Meaning of News by Etymonline," in *Online Etymology Dictionary*, accessed January 10, 2023, https://www.etymonline.com/word/news.

2   Joshua J. Mark, "The Ancient City," World History Encyclopedia, April 5, 2014, https://www.worldhistory.org/city/.

3   Jack Weatherford, *Genghis Khan and the Making of the Modern World* (New York: Crown, 2005); Wuyun Gaowa, "Yuan's Postal System Facilitated East-West Trade, Cultural Exchange-SSCP," *Chinese Social Science Today* (blog), April 6, 2017, http://www.csstoday.com/Item/4326.aspx.

4   "Postal System," in *Britannica*, accessed January 11, 2023, https://www.britannica.com/topic/postal-system/History#ref367055.

5   Andrew Pettegree, *The Invention of News: How the World Came to Know about Itself* (New Haven, CT: Yale University Press, 2014), 169.

6   Ibid., 167–81.

7   Jennifer Spinks, *Monstrous Births and Visual Culture in Sixteenth-Century Germany* (London: Pickering & Chatto, 2014), 59–79, https://www.cambridge.org/core/books/monstrous-births-and-visual-culture-in-sixteenthcentury-germany/97EE04D856A7BDCC671DB701205C4C0C; Renée DiResta and Tobias Rose-Stockwell, "How to Stop Misinformation before It Gets Shared," *Wired*, March 26, 2021, https://www.wired.com/story/how-to-stop-misinformation-before-it-gets-shared/.

8   Pettegree, 254.

9   Ibid., 251.

10 Robert H. Knapp, "A Psychology of Rumor," *Public Opinion Quarterly* 8, no. 1 (1944): 22–37.

11 Pettegree, 22.

12 Tobias Rose-Stockwell, "This Is How Your Fear and Outrage Are Being Sold for Profit," *Medium* (blog), August 12, 2019, https://tobiasrose.medium.com/the-enemy-in-our-feeds-e86511488de.

13 DiResta and Rose-Stockwell, 2021.

14 Cleveland Ferguson III, "Yellow Journalism," The First Amendment Encyclopedia, 2009, https://www.mtsu.edu/first-amendment/article/1253/yellow-journalism; Seymour Topping, "History of The Pulitzer Prizes," The Pulitzer Prizes, accessed January 10, 2023, https://www.pulitzer.org/page/history-pulitzer-prizes.

15 Kristine A. Oswald, "Mass Media and the Transformation of American Politics," *Marquette Law Review* 77 (2009).

16 Tim Wu, *The Attention Merchants: The Epic Scramble to Get inside Our Heads* (New York: Alfred A. Knopf, 2016), 175.

17 Shannon K. McCraw, "Equal Time Rule," The First Amendment Encyclopedia, 2009, https://www.mtsu.edu/first-amendment/article/949/equal-time-rule.

18 Jonathan Rauch, *The Constitution of Knowledge: A Defense of Truth* (Washington, DC: Brookings Institution Press, 2021), 95–117.

19 Angelo Fichera Spencer Saranac Hale, "Bogus Theory Claims Supercomputer Switched Votes in Election," *FactCheck.Org* (blog), November 13, 2020, https://www.factcheck

.org/2020/11/bogus-theory-claims-supercomputer-switched-votes-in-election/; Mikki Willis, *Plandemic: Fear Is the Virus. Truth Is the Cure* (New York: Skyhorse, 2021); Jessica Contrera, "A QAnon Con: How the Viral Wayfair Sex Trafficking Lie Hurt Real Kids," *Washington Post,* December 16, 2021, https://www.washingtonpost.com /dc-md-va/interactive/2021/wayfair-qanon-sex-trafficking-conspiracy/; Shahin Nazar and Toine Pieters, "Plandemic Revisited: A Product of Planned Disinformation Ampli- fying the COVID-19 'Infodemic,'" *Frontiers in Public Health* 9 (2021), https://www .frontiersin.org/articles/10.3389/fpubh.2021.649930.

20 Michael Schudson, *Discovering the News: A Social History of American Newspapers,* Nachdr. (New York: Basic Books, 1981), 39–41.

Chapter 17: The First Twitter Thread

1. Joshua J. Mark, "Johann Tetzel," World History Encyclopedia, July 28, 2022, https:// www.worldhistory.org/Johann_Tetzel/.

2 Martin Luther, *Martin Luther's 95 Theses,* ed. Stephen J. Nichols (Phillipsburg, NJ: P & R Publishing, 2002); Andrew Pettegree, *The Invention of News: How the World Came to Know about Itself* (New Haven, CT: Yale University Press, 2014), 67–72.

3 "John Wycliffe: Translation of the Bible," in *Britannica,* accessed January 10, 2023, https://www.britannica.com/biography/John-Wycliffe/Translation-of-the-Bible.

4 Elizabeth L. Eisenstein, *The Printing Press as an Agent of Change: Communications and Cultural Trans* (Cambridge, UK: Cambridge University Press, 1982); Ch'on Hye-bong, "Typography in Korea: Birthplace of Moveable Metal Type," *Korea Journal* 3, no. 7 (July 1993): 10–19.

5 Pettegree, 67–72. The Editors, "Martin Luther's Life and Legacy," in *Britannica,* accessed January 10, 2023, https://www.britannica.com/summary/Martin-Luther.

6 Martin Luther, "Against the Robbing and Murdering Hordes of Peasants," in *Essential Luther,* trans. Tryntje Helfferich (Indianapolis: Hackett, 2018).

7 "Persecution," UK Parliament, accessed January 11, 2023, https://www.parliament.uk /about/living-heritage/transformingsociety/private-lives/religion/overview/persecution/.

8 Christopher Hill, *Antichrist in Seventeenth-Century England,* rev. ed. (London: Verso, 1990) 33–39.

Chapter 18: American Outrage

1 Walter Isaacson, *Benjamin Franklin: An American Life* (New York: Simon & Schuster, 2003), 8.

2 Ibid., 227.

3 Mark Hailwood, "'The Rabble That Cannot Read'? Ordinary People's Literacy in Seventeenth-Century England," *The Many-Headed Monster* (blog), October 13, 2014, https://manyheadedmonster.com/2014/10/13/the-rabble-that-cannot-read-ordinary -peoples-literacy-in-seventeenth-century-england/.

4 John R. Vile, "John Peter Zenger," The First Amendment Encyclopedia, 2009, https:// www.mtsu.edu/first-amendment/article/1235/john-peter-zenger.

5   "The Seven Years' War," *The American Revolution Institute* (blog), July 24, 2022, https://
    www.americanrevolutioninstitute.org/video/the-seven-years-war/.

6   Chester E. Jorgenson, "The New Science in the Almanacs of Ames and Franklin,"
    New England Quarterly 8, no. 4 (1935): 555–61, https://doi.org/10.2307/360361; Den-
    nis Landis et al., "Pamphlet Wars," John Carter Brown Library, https://www.brown
    .edu/Facilities/John_Carter_Brown_Library/exhibitions/pamphletWars/pages/crisis
    .html.

7   Andrew Pettegree, *The Invention of News: How the World Came to Know about Itself*
    (New Haven, CT: Yale University Press, 2014), 334.

8   Editors, "The Tombstone Edition: Pennsylvania Journal, October 31, 1765," Journal
    of the American Revolution, June 15, 2015, https://allthingsliberty.com/2015/06/the
    -tombstone-edition-pennsylvania-journal-october-31-1765/.

9   Pettegree, 334.

10 Christopher Klein, "The Stamp Act Riots," HISTORY, August 31, 2018, https://www
    .history.com/news/the-stamp-act-riots-250-years-ago.

11 "Stamp Act | History, Definition, Facts, & Riots," in *Britannica*, December 9, 2022,
    https://www.britannica.com/event/Stamp-Act-Great-Britain-1765.

12 History.com Editors, "Boston Massacre," HISTORY, September 20, 2022, https://
    www.history.com/topics/american-revolution/boston-massacre.

13 Jayne E. Triber, *A True Republican: The Life of Paul Revere* (Amherst, MA: University of
    Massachusetts Press, 2001), 80.

14 John Adams, "Argument for the Defense: 3–4 December 1770," Founders Online,
    National Archives, http://founders.archives.gov/documents/Adams/05-03-02-0001
    -0004-0016.

15 "Boston Massacre Trial," Boston National Historical Park, U.S. National Park Service,
    February 26, 2015, https://www.nps.gov/bost/learn/historyculture/massacre-trial.htm.

Chapter 19: How Advertising Created Newspapers

1   Note that many anecdotes in this chapter come from these two fantastic books: Michael
    Schudson, *Discovering the News: A Social History of American Newspapers*, Nachdr. (New
    York: Basic Books, 1981); David T. Z. Mindich, *Just the Facts: How "Objectivity" Came
    to Define American Journalism* (New York: New York University Press, 1998).

2   Andrew Belonsky, "How the Penny Press Brought Great Journalism to Populist
    America," *Daily Beast*, September 8, 2018, https://www.thedailybeast.com/how-the
    -penny-press-brought-great-journalism-to-populist-america.

3   David W. Bulla, "Party Press Era | United States History," in *Britannica*, December
    2015, https://www.britannica.com/topic/party-press-era.

4   Isaac Clarke Pray, *Memoirs of James Gordon Bennett and His Times* (Stringer &
    Townsend, 1855), 84.

5   Robert McNamara, "1836 Murder of a New York Prostitute Changed American Journal-
    ism," ThoughtCo, July 31, 2018, https://www.thoughtco.com/murder-of-helen-jewett
    -1773772.

6   Katherine Roeder, *Wide Awake in Slumberland: Fantasy, Mass Culture, and Modernism in the Art of Winsor McCay*, Apple Books Edition (Jackson, MS: University Press of Mississippi, 2014), 48.

7   Schudson, 20.

8   Ibid., 19–20.

9   *Understanding Media and Culture*, University of Minnesota Libraries Publishing Edition (2016), 161–62, https://doi.org/10.24926/8668.2601. This edition is adapted from a work originally produced in 2010 by a publisher who has requested that it not receive attribution.

10 Schudson, 12–60.

11 Hannah Arendt and Margaret Canovan, *The Human Condition*, 2nd ed. (Chicago: University of Chicago Press, 1998), 35.

12 Dan Schiller, *Objectivity and the News: The Public and the Rise of Commercial Journalism* (Philadelphia: University of Pennsylvania Press, 1981), 76–79; Meg Matthias, "The Great Moon Hoax of 1835 Was Sci-Fi Passed Off as News," in *Britannica*, accessed January 10, 2023, https://www.britannica.com/story/the-great-moon-hoax-of-1835-was-sci-fi-passed-off-as-news.

13 Roger Fenton, "Roger Fenton (1819–69): Valley of the Shadow of Death," Royal Collection Trust, accessed January 10, 2023, https://www.rct.uk/collection/2500514/valley-of-the-shadow-of-death.

14 Kathryn Schulz, "Errol Morris Looks for the Truth in Photography," *New York Times*, September 1, 2011, sec. Books, https://www.nytimes.com/2011/09/04/books/review/believing-is-seeing-by-errol-morris-book-review.html.

15 Schiller, 76–79.

16 Karl E. Meyer, "150th Anniversary: 1851–2001; Dept. of Conscience: The Editorial 'We,' " *New York Times*, November 14, 2001, sec. Archives, https://www.nytimes.com/2001/11/14/news/150th-anniversary-1851-2001-dept-of-conscience-the-editorial-we.html.

17 Clarence Darrow, *Realism in Literature and Art* (Girard, KS: Haldeman-Julius Company, 1899), 21.

18 Robert Hoe, *A Short History of the Printing Press and of the Improvements in Printing Machinery from the Time of Gutenberg up to the Present Day* (Alpha Edition, 2020).

Chapter 20: The Dark Valley of Radio

1   Donald Warren, *Radio Priest: Charles Coughlin, The Father of Hate Radio* (New York: Free Press, 1996); Charles E. Coughlin, *Father Coughlin's Radio Discourses 1931–1932* (Cabin John, MD: Wildside Press, 2021).

2   Oliver Rathkolb, *Revisiting the National Socialist Legacy: Coming to Terms With Forced Labor, Expropriation, Compensation, and Restitution* (Piscataway, NJ: Transaction Publishers, 2004), 82; Allison C. Meier, "An Affordable Radio Brought Nazi Propaganda Home," *JSTOR Daily*, August 30, 2018, https://daily.jstor.org/an-affordable-radio-brought-nazi-propaganda-home/.

3   Rathkolb, 82.

4   "Charles Coughlin—Americans and the Holocaust," United States Holocaust Memorial Museum, accessed January 10, 2023, https://exhibitions.ushmm.org/americans-and-the-holocaust/personal-story/charles-coughlin.

5   Thomas Doherty, "The Deplatforming of Father Coughlin," *Slate*, January 21, 2021, https://slate.com/technology/2021/01/father-coughlin-deplatforming-radio-social-media.html.

6   Renée DiResta, "Free Speech Is Not the Same as Free Reach," *Wired*, accessed January 10, 2023, https://www.wired.com/story/free-speech-is-not-the-same-as-free-reach/. This phrase was popularized by writer Renée DiResta, paraphrasing Aza Raskin.

Chapter 21: Television, Chaos, and the Collective

1   Bryan Burrough, *Days of Rage: America's Radical Underground, the FBI, and the Forgotten Age of Revolutionary Violence*, Apple Books Edition, repr. ed. (New York: Penguin Books, 2016), 25.

2   Peter Bell, "Public Trust in Government: 1958–2022," *U.S. Politics & Policy* (blog), Pew Research Center, June 6, 2022, https://www.pewresearch.org/politics/2022/06/06/public-trust-in-government-1958-2022/.

3   Sarah Sobieraj and Jeffrey M. Berry, "From Incivility to Outrage: Political Discourse in Blogs, Talk Radio, and Cable News," *Political Communication* 28, no. 1 (February 9, 2011): 19–41, https://doi.org/10.1080/10584609.2010.542360.

4   *Network*, Drama (Metro-Goldwyn-Mayer, 1976).

5   "Fairness Doctrine," Ronald Reagan Presidential Library & Museum, December 16, 2021, https://www.reaganlibrary.gov/archives/topic-guide/fairness-doctrine.

6   "Fairness Doctrine."

7   David Swistock, John Nielsen, and Devin Gillen, "Rush Limbaugh," *History in the Making* 14, no. 1 (July 29, 2021), https://scholarworks.lib.csusb.edu/history-in-the-making/vol14/iss1/13.

8   Jeremy W. Peters, "Rush Limbaugh's Legacy of Venom: As Trump Rose, 'It All Sounded Familiar,'" *New York Times*, February 18, 2021, sec. U.S., https://www.nytimes.com/2021/02/17/us/politics/limbaugh-death.html.

9   Jeremy Barr, "The Downside of Being a Fox News Journalist? Getting Asked about Sean Hannity," *Hollywood Reporter*, May 8, 2020, https://www.hollywoodreporter.com/tv/tv-news/downside-being-a-fox-news-journalist-getting-asked-sean-hannity-1292957/.

10   Yochai Benkler, Robert Farris, and Hal Roberts, *Network Propaganda: Manipulation, Disinformation, and Radicalization in American Politics* (New York: Oxford University Press, 2018), https://doi.org/10.1093/oso/9780190923624.001.0001.

Chapter 22: How We Learn the Truth

1   Michael Schudson, *Discovering the News: A Social History of American Newspapers*, Nachdr. (New York: Basic Books, 1981) 92.

2   "'Yellow Journalism,'" Evening Standard, September 25, 1901, http://mckinleydeath.com/documents/newspapers/EStandard092501a.htm; Louis Anslow, "Before the

Internet, Irresponsible Journalism Was Blamed for a War and a Presidential Assassination," *Medium*, February 9, 2017, https://timeline.com/yellow-journalism-media-history-8a 29e4462ac.

3   Bill Kovach and Tom Rosenstiel, *The Elements of Journalism*, 4th rev. ed. (New York: Crown, 2021), 101, 121.

4   Alexander Hamilton, Donald R. Hickey, and Connie D. Clark, *Citizen Hamilton: The Wit and Wisdom of an American Founder* (Lanham, MD: Rowman & Littlefield, 2006), 112.

5   Amy Mitchell, "Distinguishing Between Factual and Opinion Statements in the News," *Pew Research Center's Journalism Project* (blog), June 18, 2018, https://www.pewresearch.org/journalism/2018/06/18/distinguishing-between-factual-and-opinion-statements-in-the-news/.

6   Mollie Leavitt, "Q&A: James Hamilton, Director of Stanford University's Journalism Program," *The Idea* (blog), August 19, 2019, https://medium.com/the-idea/q-a-james-hamilton-director-of-stanford-universitys-journalism-program-779a87486edf.

7   Farida B. Ahmad and Robert N. Anderson, "The Leading Causes of Death in the US for 2020," *JAMA* 325, no. 18 (May 11, 2021): 1829–30, https://doi.org/10.1001/jama.2021.5469.

8   Nemil Dalal, "Today's Biggest Threat to Democracy Isn't Fake News—It's Selective Facts," *Quartz*, November 16, 2017, https://qz.com/1130094/todays-biggest-threat-to-democracy-isnt-fake-news-its-selective-facts/.

9   Carol M. Liebler, "Me(Di)a Culpa?: The 'Missing White Woman Syndrome' and Media Self-Critique," *Communication, Culture & Critique* 3, no. 4 (2010): 549–65, https://doi.org/10.1111/j.1753-9137.2010.01085.x.

10 Logan Molyneux, personal correspondence, Zoom, September 30, 2021.

11 "Network Effects," *Economist*, September 24, 2013, https://www.economist.com/christmas-specials/2013/09/24/network-effects.

12 Edward S. Herman and Noam Chomsky, *Manufacturing Consent: The Political Economy of the Mass Media* (New York: Knopf Doubleday Publishing Group, 2011).

13 *Camel News Caravan—19/September/1952*, Internet Archive, http://archive.org/details/CamelNewsCaravan-19september1952.

14 Paul Virilio, *The Original Accident*, trans. Julie Rose, 1st edition (Cambridge, UK: Polity, 2007).

15 New York–based media executive, personal correspondence, June 17, 2021.

16 Daniel Trotta, "Iraq War Costs U.S. More than $2 Trillion: Study," *Reuters*, March 14, 2013, sec. U.S. Markets, https://www.reuters.com/article/us-iraq-war-anniversary-idUSBRE92D0PG20130314; David Vine et al., "Millions Displaced by U.S. Post-9/11 Wars" (Watson Institute, Brown University, August 19, 2021), https://watson.brown.edu/costsofwar/files/cow/imce/papers/2021/Costs%20of%20War_Vine%20et%20al_Displacement%20Update%20August%202021.pdf.

17 Jonathan Rauch, *The Constitution of Knowledge: A Defense of Truth* (Washington, DC: Brookings Institution Press, 2021), 121.

18 Thomas Stockwell, private interview, April 2020.

19   Jeremy Littau, "Why Do All These Media Layoffs Keep Happening? A Thread,"
     Medium (blog), February 4, 2019, https://jeremylittau.medium.com/why-do-all-these
     -media-layoffs-keep-happening-a-thread-34b4b4edbe8c.
20   Jeremy Littau, "The Crisis Facing American Journalism Did Not Start with the Inter-
     net," *Slate*, January 26, 2019, https://slate.com/technology/2019/01/layoffs-at-media
     -organizations-the-roots-of-this-crisis-go-back-decades.html.
21   David Streitfeld, "Craig Newmark, Newspaper Villain, Is Working to Save Journalism,"
     New York Times, October 17, 2018, https://www.nytimes.com/2018/10/17/technology
     /craig-newmark-journalism-gifts.html.

Chapter 23: Trust and Truth

1   David McRaney, "How to Improve Your Chances of Nudging the Vaccine Hesitant
     Away from Hesitancy and toward Vaccination," *You Are Not So Smart Podcast*, https://
     youarenotsosmart.com/2021/08/23/yanss-213-how-to-improve-your-chances-of-nudg
     ing-the-vaccine-hesitant-away-from-hesitancy-and-toward-vaccination/.
2   Jay J. Van Bavel and Andrea Pereira, "The Partisan Brain: An Identity-Based Model of
     Political Belief," *Trends in Cognitive Sciences* 22, no. 3 (March 1, 2018): 213–24, https://doi
     .org/10.1016/j.tics.2018.01.004.
3   Susan T. Fiske and Shelley E. Taylor, *Social Cognition: From Brains to Culture* (Thousand
     Oaks, CA: SAGE, 2013).
4   Scott Alexander, "Ivermectin: Much More Than You Wanted to Know," Substack
     newsletter, *Astral Codex Ten* (blog), November 17, 2021, https://astralcodexten.substack
     .com/p/ivermectin-much-more-than-you-wanted.
5   Laura A. Kurpiers et al., "Bushmeat and Emerging Infectious Diseases: Lessons from
     Africa," in *Problematic Wildlife: A Cross-Disciplinary Approach*, ed. Francesco M. Angelici
     (Cham, Switzerland: Springer International Publishing, 2016), 507–51, https://doi
     .org/10.1007/978-3-319-22246-2_24; Ben Westcott and Serenetie Wang, "China's
     Wet Markets Are Not What Some People Think They Are," CNN World, April 23,
     2020, https://www.cnn.com/2020/04/14/asia/china-wet-market-coronavirus-intl-hnk
     /index.html.
6   Glenn Kessler, "Analysis | Timeline: How the Wuhan Lab-Leak Theory Suddenly
     Became Credible," *Washington Post*, May 27, 2021, https://www.washingtonpost.com
     /politics/2021/05/25/timeline-how-wuhan-lab-leak-theory-suddenly-became-credible/.
7   CDC (@CDCgov), "CDC does not currently recommend the use of facemasks to
     help prevent novel #coronavirus. Take everyday preventive actions, like staying home
     when you are sick and washing hands with soap and water, to help slow the spread
     of respiratory illness. #COVID19," Tweet, *Twitter*, February 27, 2020, https://twit-
     ter.com/CDCgov/status/1233134710638825473; Anna Hecht, "These 3 Etsy Shop
     Owners Have Each Sold Hundreds of Cloth Face Masks Since the Pandemic Started,"
     CNBC Make It, May 7, 2020, https://www.cnbc.com/2020/05/07/etsy-shop-owners
     -sell-hundreds-of-cloth-face-masks-during-pandemic.html.

Chapter 24: Freedom of Speech vs. Defense of Truth

1 Aaron Couch, "Jon Stewart and Bill O'Reilly Get into Shouting Match over 'White Privilege,'" *Hollywood Reporter*, October 15, 2014, https://www.hollywoodreporter .com/tv/tv-news/jon-stewart-bill-oreilly-get-741276/.

2 "Special Video Reports—The Private Lives Of George Washington's Slaves," *Frontline PBS*, https://www.pbs.org/wgbh/pages/frontline/shows/jefferson/video/lives.html.

3 Andrew Roberts, "Book Review: Great Soul," *Wall Street Journal*, accessed January 11, 2023, https://www.wsj.com/articles/SB10001424052748703529004576160371482469358.

4 John Stuart Mill and Elizabeth Rapaport, *On Liberty* (Indianapolis: Hackett, 1978).

5 Mill and Rapaport, 18.

6 Ibid., 16–35.

7 Karl R. Popper, Alan Ryan, and E. H. Gombrich, *The Open Society and Its Enemies: New One-Volume Edition* (Princeton, NJ: Princeton University Press, 2013), 581.

8 John Rawls, *A Theory of Justice*, rev. ed. (Cambridge, MA: Belknap Press, 1999).

9 Thomas Jefferson, "First Inaugural Address," in *The Papers of Thomas Jefferson, 17 February to 30 April 1801*, vol. 33 (Princeton, NJ: Princeton University Press, 2006), 148–52, https://jeffersonpapers.princeton.edu/selected-documents/first-inaugural-address-0.

10 Andrew Pettegree, *The Invention of News: How the World Came to Know about Itself* (New Haven, CT: Yale University Press, 2014), 369.

11 Robert McNamara, "Abolitionist Pamphlets Sent to the South Sparked Controversy," ThoughtCo, January 31, 2020, https://www.thoughtco.com/abolitionist-pamphlet -campaign-1773556.

12 Nadine Strossen, *HATE: Why We Should Resist It With Free Speech, Not Censorship* (New York: Oxford University Press, 2018); Zack Beauchamp, "Why Book Banning Is Back in 2022," *Vox*, February 10, 2022, https://www.vox.com/policy-and-politics/22914767 /book-banning-crt-school-boards-republicans.

13 Jonathan Rauch, *The Constitution of Knowledge: A Defense of Truth* (Washington, DC: Brookings Institution Press, 2021), 35.

Chapter 25: The Parable of the Island

1 Brian Stelter, "This Infamous Steve Bannon Quote Is Key to Understanding America's Crazy Politics," *CNN Business*, November 16, 2021, https://www.cnn.com/2021/11 /16/media/steve-bannon-reliable-sources/index.html.

Chapter 26: What's at Stake

1 Abigail Geiger, "Political Polarization in the American Public," *U.S. Politics & Policy* (blog), Pew Research Center, June 12, 2014, https://www.pewresearch.org /politics/2014/06/12/political-polarization-in-the-american-public/.

2 Henry E. Hale, "25 Years after the USSR: What's Gone Wrong?" *Journal of Democracy* 27, no. 3 (2016): 9, https://doi.org/10.1353/jod.2016.0035.

3 Francis Fukuyama, "The End of History?" *National Interest*, no. 16 (1989): 3–18.

4   Jorge I. Domínguez, "Boundary Disputes in Latin America" (United States Institute for Peace, September 2003), https://www.usip.org/sites/default/files/resources/pwks50.pdf; Laust Schouenborg, "Why War Has Become Obsolete in Europe," Spice Stanford, 2010, http://spice.fsi.stanford.edu/docs/why_war_has_become_obsolete_in_europe; Muthiah Alagappa, "International Peace in Asia: Will It Endure?" Carnegie Endowment for International Peace, December 19, 2014, https://carnegieendowment.org/2014/12/19/international-peace-in-asia-will-it-endure-pub-57588.

5   Jean-Jacques Rousseau, *The Social Contract* (CreateSpace Independent Publishing Platform, 2014).

6   Thomas Jefferson, *The Declaration of Independence and the Constitution of the United States: With Index*, ed. Pauline Maier (New York: Bantam Books, 1998), 53.

7   John Locke, *The Second Treatise on Civil Government*, Great Books in Philosophy Edition (New York: Prometheus, 1986); Thomas Hobbes, *Leviathan* (Touchstone, 1997), 78, 80; Robb A. McDaniel, "John Locke," The First Amendment Encyclopedia, accessed January 11, 2023, https://www.mtsu.edu/first-amendment/article/1257/john-locke.

8   Karen Stenner, *The Authoritarian Dynamic, Cambridge Studies in Public Opinion and Political Psychology* (New York: Cambridge University Press, 2005), 330.

9   Sarah Repucci and Amy Slipowitz, *Democracy under Siege* (Washington, DC: Freedom House, 2021), https://freedomhouse.org/report/freedom-world/2021/democracy-under-siege; "The Rise and Risks of 'The Age of the Strongman,'" *Economist*, April 9, 2022, https://www.economist.com/culture/2022/04/09/the-rise-and-risks-of-the-age-of-the-strongman.

Chapter 27: The Machine Called Democracy

1   Alexander Hamilton, James Madison, and John Jay, eds., "No. 55: The Total Number of the House of Representatives," in *The Federalist Papers*, Dover Thrift Editions (Mineola, NY: Dover Publications, Inc, 2014), 272.

2   Alexander Hamilton, James Madison, and John Jay, eds., "No. 10: The Same Subject Continued," in *The Federalist Papers*, Dover Thrift Editions (Mineola, NY: Dover Publications, Inc, 2014), 47.

3   Robert Cohen, "Was the Constitution Pro-Slavery? The Changing View of Frederick Douglass," *Social Education* (2008), 246–50. Damon Root, "When the Constitution Was 'at War with Itself,' Frederick Douglass Fought on the Side of Freedom," *Reason Magazine*, February 2, 2018, https://reason.com/2018/02/02/when-the-constitution-was-at-war-with-it/.

4   Neil Harvey, "Toward a More Perfect Union: Unleashing the Promise in Us All with Angela Glover Blackwell," *Bioneers: Revolution From the Heart of Nature*, accessed January 9, 2023, https://bioneers.org/toward-a-more-perfect-union-unleashing-the-promise-in-us-all-with-angela-glover-blackwell/.

5   A. H. Land and A. G. Doig, "An Automatic Method of Solving Discrete Programming Problems," *Econometrica* 28, no. 3 (1960): 497–520, https://doi.org/10.2307/1910129.

6   Daniel Crofts, "Communication Breakdown," *New York Times Opinionator*, May 21, 2011, sec. Opinion, https://archive.nytimes.com/opinionator.blogs.nytimes.com/2011/05/21/communication-breakdown/.

7   Annika Neklason, "The Conspiracy Theories That Fueled the Civil War," *Atlantic*, May 29, 2020, https://www.theatlantic.com/politics/archive/2020/05/conspiracy-theories-civil -war/612283/.

Chapter 28: Where Should We Place Our Outrage?

1   Allen Ginsberg, *Howl: And Other Poems*, Nachdr., The Pocket Poets Series 4 (San Francisco: City Lights Books, 2010), 17.
2   Scott Alexander, "Meditations On Moloch | Slate Star Codex," *Slate Star Codex* (blog), July 30, 2014, https://slatestarcodex.com/2014/07/30/meditations-on-moloch/.
3   Alexander.
4   Robert Axelrod and William D. Hamilton, "The Evolution of Cooperation," *Science* 211, no. 4489 (March 1981): 1390–6, https://doi.org/10.1126/science.7466396. Nicky Case, "The Evolution of Trust," July 2017, http://ncase.me/trust/.

Chapter 29: What You Can Do

1   Hans Rosling, Anna Rosling Rönnlund, and Ola Rosling, *Factfulness: Ten Reasons We're Wrong about the World—and Why Things Are Better Than You Think*, later print ed. (New York: Flatiron Books, 2018), 75.
2   The nonpartisan, nonprofit Constructive Dialogue Institute is a great resource for having conversations like these. https://constructivedialogue.org/.
3   Joe Pinsker, "Trump's Presidency Is Over. So Are Many Relationships," *The Atlantic*, March 30, 2021, https://www.theatlantic.com/family/archive/2021/03/trump-friend -family-relationships/618457/.